Gustav Robert Kirchhoff's Treatise
"On the Theory of Light Rays" (1882)

English Translation, Analysis and Commentary

Gustav Robert Kirchhoff's Treatise
"On the Theory of Light Rays" (1882)

English Translation, Analysis and Commentary

Editors

Klaus Hentschel
Ning Yan Zhu

University of Stuttgart, Germany

World Scientific

NEW JERSEY • LONDON • SINGAPORE • BEIJING • SHANGHAI • HONG KONG • TAIPEI • CHENNAI • TOKYO

Published by

World Scientific Publishing Co. Pte. Ltd.

5 Toh Tuck Link, Singapore 596224

USA office: 27 Warren Street, Suite 401-402, Hackensack, NJ 07601

UK office: 57 Shelton Street, Covent Garden, London WC2H 9HE

Library of Congress Cataloging-in-Publication Data

Names: Kirchhoff, G. (Gustav), 1824–1887, author. | Hentschel, Klaus, editor. | Zhu, Ning Yan,
 editor. | Kirchhoff, G. (Gustav), 1824–1887. Zur Theorie der Lichtstrahlen. English.
Title: Gustav Robert Kirchhoff's treatise "On the theory of light rays" (1882) :
 English translation, analysis and commentary / edited by Klaus Hentschel
 (University of Stuttgart, Germany), Ning Yan Zhu (University of Stuttgart, Germany).
Description: Hackensack, NJ : World Scientific, [2016] | Includes index.
Identifiers: LCCN 2016024822| ISBN 9789813147133 (hardcover ; alk. paper) |
 ISBN 981314713X (hardcover ; alk. paper) | ISBN 9789813147140 (pbk. ; alk. paper) |
 ISBN 9813147148 (pbk. ; alk. paper)
Subjects: LCSH: Diffraction. | Spectrum analysis. | Kirchhoff, G. (Gustav), 1824–1887.
Classification: LCC QC415 .K57 2016 | DDC 535.01--dc23
LC record available at https://lccn.loc.gov/2016024822

British Library Cataloguing-in-Publication Data
A catalogue record for this book is available from the British Library.

Typeset by Stallion Press
Email: enquiries@stallionpress.com

Editors' Foreword

The idea for this volume goes back to an open day arranged once a year by the University of Stuttgart. The booths of history of science and technology, and of electrical engineering happened to be next to each other, and that's how we two editors first met. We soon realized that we both have a strong interest in the history of electrodynamics. Ning Yan Zhu mentioned an old paper from 1882 by the mathematical physicist Kirchhoff on diffraction theory, lamenting that it had never been translated into English. This meant that many of his colleagues — and Anglo-Saxon historians of science as well — had, in fact, never read the paper. Since the wife of one of the editors is a professional translator specializing in history of science, it was straightforward to arrange that Ann Hentschel translate the Kirchhoff treatise. Klaus Hentschel had already written various dictionary and encyclopedia entries on Gustav Robert Kirchhoff and his scientific achievements, particularly in spectroscopy and spectrum analysis, so it was also straightforward that he would contribute the biographical introduction to the volume. Ning Yan Zhu wrote a brief commentary to the primary text in order to guide the reader through the thicket of formulas in Kirchhoff's quite challenging text. Further enrichment to the planned volume arose when Klaus Hentschel spoke to a bibliometrician at the library unit of the nearby *Max Planck Institute for Solid State Research* in Stuttgart-Büsnau. Werner Marx specializes in bibliometric analyses of older historical literature and thus offered to study the reception of Kirchhoff's paper through the ages. The two last additions came from farther afield. Peter Vickers from *Durham University* in England has a long-standing interest in Kirchhoff's paper as a paradigmatic case for counterfactual assumptions. Why Kirchhoff's approximation works nevertheless is the topic of his contribution. Jed Z. Buchwald at *Caltech*, Pasadena, California, and Chen-Pang

Yeang at the *Institute for the History and Philosophy of Science and Technology* in Toronto, Canada, provided their own views on Kirchhoff's theory of optical diffraction, its predecessor and subsequent development. So now we have in fact collected three different, in some points conflicting, but certainly also mutually embellishing views on the reasons for the resilience of an inconsistent theory. We hope that this anthology will make all English-speaking scientists, engineers, historians, and interested laymen aware of the great fecundity of Kirchhoff's thought and historical context.

The paper by Jed Z. Buchwald and Chen-Pang Yeang also appeared in the *Archive for History of Exact Sciences* in 2016. We appreciate the permission granted by *Springer International Publishing AG* to republish it here. It was a great pleasure to work with Andrea Wolf in the Munich office of *World Scientific* and her colleagues in their international production division.

University of Stuttgart, March 2016
Prof. Dr. Klaus Hentschel
(Section for History of Science and Technology) and
Dr. Ning Yan Zhu (Institute of Radio Frequency Technology)

Contents

Editors' Foreword   v

Biographical Introduction, by Klaus Hentschel   1

Commentary on Kirchhoff's Theory of Diffraction,
     by Ning Yan Zhu   19

Gustav Robert Kirchhoff: *On the Theory of Light Rays*   31

Kirchhoff's Theory for Optical Diffraction,
     by Jed Z. Buchwald and Chen-Pang Yeang   63

Why Kirchhoff's Approximation Works, by Peter Vickers   125

A Brief Bibliometric Analysis by Werner Marx   143

Name Index   153

List of Figures

1.   Photograph of young Gustav Robert Kirchhoff c. 1860   2

2.   Bunsen's and Kirchhoff's first spectroscope 1859   8

3.   Photograph of Kirchhoff, Bunsen and Roscoe in 1862   9

4.   Portrait of the elder Gustav Robert Kirchhoff c. 1880   17

5.   Fresnel's configuration for diffraction   68

6.   Stokes' configuration for diffraction   74

7.   Kirchhoff's configuration for Green's theorem   82

8.   Kirchhoff's further configuration for Green's theorem   85

9.   Kirchhoff's diffraction configuration   89

10. Poincaré's first surfaces   92

11. Poincaré's surfaces adapted to an opaque screen   94

12. Poincaré's point L within the bounded region   95

13. Sommerfeld's configuration for his Green's function   102

14. Rubinowicz's surfaces   108

15. A birds-eye view of the theoretical situation   131

16. Kirchhoff's amplitude function and Maxwell's equations   133

17. Behaviour of the light at the aperture   134

18. the polarisation case   135

19. Polarisation   138

20. Time-dependent number of formal and informal citations . . .   146

21. Citation history   147

Biographical Introduction, by Klaus Hentschel

Kirchhoff the Classical Scientist

Nowadays Gustav Robert Kirchhoff[1] is primarily known for his laws about the distribution of electric currents in a network and as co-discoverer of spectrum analysis in 1860, one of the most important areas of scientific research during the nineteenth century. By comparison, his individual analyses on divers problems in mathematics and theoretical physics are familiar only to specialists. His textbook on mathematical physics earned him renown for establishing theoretical physics as an independent discipline in Germany.[2] Many leading theoreticians of the next generation profited from his instruction. They include Heinrich Hertz, Eilhart Wiedemann, Erich Bessel-Hagen and Max Planck, as well as Ludwig Boltzmann and Victor von Lang from Austria, the Luxembourg-born Gabriel Jonas Lippmann from France, Heike Kamerlingh-Onnes from the Netherlands and the native German Arthur Schuster from England. They particularly appreciated their teacher's "extreme care and conscientiousness." Kirchhoff is an excellent example of what Wilhelm Ostwald typified as the 'classical' scientist. Although he published relatively little,

---

[1]The most detailed biography is by Kurt Hübner: *Gustav Robert Kirchhoff*, Ubstadt-Weiher: Verlag Regionalkultur, 2010; cf. also Klaus Hentschel: Gustav Robert Kirchhoff und seine Zusammenarbeit mit Robert Wilhelm Bunsen, in: Karl v. Meyenn (ed.) *Die Großen Physiker*, Munich: Beck, vol. 1 (1997), pp. 416–430, 475–477, 532–534 (both only available in German).

[2]Gustav Robert Kirchhoff: *Vorlesungen über mathematische Physik*, Leipzig: Teubner, vol. 1: *Mechanik* 1876, vol. 2: *Mathematische Optik*, K. Hensel (ed.) 1891. Vol. 3: *Elektrizität und Magnetismus* 1891 and vol. 4: *Theorie der Wärme* 1894 were edited by Max Planck.

Figure 1:   Photographic portrait of young Gustav Robert Kirchhoff, dated c. 1860, from http://www.uni-heidelberg.de/institute/fak12/texte/kirchhoff.html.

almost all that he did send to press were well-rounded and carefully conceived milestones in the published literature.[3] Kirchhoff's approach became a paradigm of the phenomenological conceptualization, so typical of the period. His goal was to provide the clearest and most complete description possible of observables. By demonstrating the close relationship between the properties of emission and absorption as well as introducing the related concept of the 'black body,' Kirchhoff paved the way to quantum theory.

Kirchhoff's Laws and Research on the Theory of Electricity

Gustav Robert Kirchhoff was born in Prussia in the city of Königsberg, now Kaliningrad in Russia, on March 12th 1824. His father was a counselor

---

[3]Kirchhoff's collected works, *Gesammelte Abhandlungen*, Leipzig: Barth, 1882 fill 641 pages, and the posthumous volume, *Nachtrag*, edited by Ludwig Boltzmann in 1891 adds 137 more pages. This is little compared to the multi-volume works of many other physicists and mathematicians from the period.

of justice. After graduating from the Kneiphofsches Gymnasium in 1842, he decided to register at the local university. There he attended courses in mathematics taught by Ludwig Otto Hesse and Friedrich Julius Richelot, in astronomy by Friedrich Wilhelm Bessel and in analytical mechanics by Carl Gustav Jacob Jacobi. After his first year Kirchhoff already regarded his professor of physics Franz Ernst Neumann (1798–1895) as his most important teacher. A tradition of precision measurement had emerged under this scientist. Neumann's physics laboratory for teaching purposes was among the first in Europe.

The stamp of this Königsberg mentor is clearly perceptible in Kirchhoff's earliest research efforts. But Kirchhoff was careful not to let Neumann's quest for precision become an end in itself. His first paper, published in the *Annalen der Physik* in 1845, on the conduction of an electric current through a plane, cited no experimental margins of error whatsoever, something extremely unusual for an article coming out of the Neumann school.[4] In order to be able to compare theoretical predictions against the equipotential curves for a metal disk, Kirchhoff tested the resistance in different regions of the copper disk using branched circuits connected to it at various places. The famous 'Kirchhoff's laws' about branching current in any linear network are tucked away in a footnote at the very end of this paper. It manifests the talent of this 21-year old student in deriving simple laws in agreement with experiments from mathematical generalizations. His formulation was:

> If galvanic currents flow through a system of entirely arbitrarily interconnected wires, then:
>
> 1. if the wires $1, 2, \ldots \mu$ meet at one point,
> $I_1 + I_2 + \cdots + I_\mu = 0$, [...]
> 2. if the wires $1, 2 \ldots \nu$ form a closed shape,
> $I_1\omega_1 + I_2\omega_2 + \cdots + I_\nu\omega_\nu =$ the sum of all electromotive forces on this path: $1, 2 \ldots \nu$;

---

[4]See Gustav Robert Kirchhoff: Ueber den Durchgang eines elektrischen Stromes durch eine Ebene, insbesondere durch eine kreisförmige, *Annalen der Physik* (2nd ser.) **64** (1845): 497–514, esp. p. 509 and plate V; reprinted in his collected works: Kirchhoff 1882, p. 12. Cf. ibid. **67** (1846), pp. 344–349, **72** (1847): 497–508, **75** (1848): 189–205. An idealized curve merely plots "approximate positions of the [measured] points". Cf. Kathryn M. Olesko: *Physics as a Calling. Discipline and Practice in the Königsberg Seminar for Physics*, Ithaca and London: Cornell Univ. Press, 1991, esp. pp. 180ff. and 187f.

where $\omega_1, \omega_2 \ldots$ denote the wire resistances, $I_1, I_2 \ldots$ the intensities of the currents $[\ldots]$.[5]

In 1847 he took his doctorate with a thesis on the intensity of induced electric currents. It was inspired by a competition staged by the Philosophical Faculty and concerned the determination of the constant ϵ, "on which the intensity of induced electric currents depends." This knack for identifying phenomenological laws reemerges in his generalization of Ohm's law by relating a current's intensity I with the tension (voltage) and the external or internal resistance in a three-dimensional conductor. In 1849 Kirchhoff succeeded in unifying the mathematics of electrostatics with the theory of electric currents (in those days still called 'galvanic chains'). This problem had been outstanding since Ohm's research from 1826 and 1827. He showed that Georg Simon Ohm's (1789–1854) concept of electric tension (*Spannung*) was identifiable with electrostatic potential at least for time-independent currents.[6] In a few papers from 1857 and 1877 he derived the motion of electric charges in conductors (which Kirchhoff visualized by a fluid model) from a set of partial differential equations, attaching suitable secondary conditions.[7] From these equations he could extract the laws governing the flow of electricity within a closed circuit. He emphasized that in the limiting case of an infinitesimal electrical resistance, these differential equations were mathematically analogous to those for the dispersion of longitudinal oscillations in a vibrating rod. The opposite limiting case of infinite resistance makes the flow of electricity the analog of heat diffusion. This way of interrelating apparently disparate phenomena based on mathematical structure was typical of Kirchhoff. But as a skeptic of hypothetical elements in physics, he was wary of drawing conclusions about physical similarities based on mathematical arguments. Although his calculations of the first limiting case mentioned above led him to conclude that the propagation velocity of waves inside the rod was the same as the velocity of light, he did not extend this further to an intrinsic relation between optical waves and electric and magnetic waves. This stopping short of the breakthrough that James Clerk Maxwell and Heinrich Hertz would

---

[5]Kirchhoff 1845, pp. 513–514 or Kirchhoff 1882, pp. 15–16.

[6]Thomas Archibald: Tension and potential from Ohm to Kirchhoff, *Centaurus* **31** (1988): 141–163.

[7]Kirchhoff: Ueber die Bewegung der Elektricität in Drähten *Annalen der Physik* (2nd ser.) **100** (1857): 193–217; Ueber die Bewegung der Elektricität in Leitern, **102** (1858): 529–544.

later make toward a theory of electromagnetic waves has been regarded as a major weakness of Kirchhoff. But the epistemology underlying his approach would not allow him to take so very bold a step. His strength lay in mathematically elegant descriptions of complex phenomena, such as elastic oscillations in a circular plate.[8] In 1850 he derived their closed oscillatory modes from a single variation equation that agreed very well with measurement data, for instance, by Ernst Chladni (1756–1827).[9] The corresponding problem for non-circular plates such as the quadratic and rectangular plates which Cladni had used for his experiments with sand or fine dust on metal plates proved too difficult to solve in the 19th century and only became tractable with the development of a new mathematical approach by Walther Ritz (1878–1909) in 1902.[10] In 1859, Kirchhoff was able to apply to a free parameter, in his theory of elasticity, measurements of his own about the relation between transversal and longitudinal contractions of a hardened steel rod. Thus he was able to round off the elasticity theory attributable to Cauchy and other famous French mathematicians without having to resort to any atomistic model. Kirchhoff's theory of the motion of electricity in submerged telegraph cables from 1877 generalized the findings William Thomson (Lord Kelvin) had been offering since 1855.[11] It proved extremely useful at a time when the international telegraph cable grid was being developed on a grand scale.

Kirchhoff's Further Course

In 1847 Kirchhoff married Clara Richelot, the daughter of one of his professors at university. A year later he qualified himself for academic

---

[8] Kirchhoff: Ueber das Gleichgewicht und die Bewegung einer elastischen Scheibe, *Journal für die reine und angewandte Mathematik* **40** (1850): 51–88.

[9] See e.g., Isaac Todhunter: *A History of the Theories of Elasticity and of the Strength of Materials*, Cambridge Univ. Press: 2 vols., here vol. 2, pp. 41, 56, as well as Christa Jungnickel & Russell McCormmach: *Intellectual Mastery of Nature: Theoretical Physics from Ohm to Einstein*, Univ. of Chicago Press: 2 vols., here vol. 1, pp. 294f. On Chladni see also Myles Jackson: *Harmonious Triads: Physicists, Musicians, and Instrument Makers in Nineteenth-Century Germany*, Cambridge, Mass.: MIT Press, 2006.

[10] For more on this issue see Martin J. Gander: From Euler, Ritz and Galerkin to modern computing, *SIAM Review* 54, 4 (2012): 627–666, esp. pp. 643f.

[11] See G. Kirchhoff: Theorie der Bewegung der Elektrizität in unterseeischen und unterirdischen Telegraphendrähten (1877) and William Thomson: *Reprint of Papers on Electrostatics and Magnetism*, London: MacMillan 1884, esp. pp. 51ff.

teaching with the *Habilitation* degree in Berlin; and in late 1849 the young *Privatdozent* received his first appointment as extraordinary professor at the University of Breslau. This was where his path first crossed Robert Wilhelm Bunsen's (1811–1899).[12] Bunsen, a chemist considerably interested in applying physical methods to his field, became his colleague in Breslau in 1851 as professor of chemistry. When Bunsen left three semesters later for Heidelberg, he lobbied hard to have Kirchhoff join him in that southern city on the banks of the Neckar, as successor to the departing physics professor. Being dean of the Philosophical Faculty in 1854, Bunsen held the best possible cards. Philipp von Jolly had decided to leave because Munich had offered him far better working conditions. Moreover, Heidelberg could not attract many of the most famous physicists of the day, who included Gustav Magnus and Heinrich Wilhelm Dove in Berlin, Wilhelm Weber in Göttingen, Carl Neumann in Königsberg or Andreas von Ettingshausen in Vienna. The competition posed by Prussian universities was simply too formidable, despite the efforts by the provincial Ministry of Culture for Baden to turn the university town into the national center for studies in the sciences and medicine.[13]

At thirty years of age, Kirchhoff still had little teaching experience but he excelled as a proven researcher — von Ettinghausen already considered him to be "one of the best mathematical physicists in Germany." Bunsen peddled him as: "among the most talented younger physicists in the exact Gauß school." Weber's evaluation recommended that having Kirchhoff at the same location as Bunsen could only be mutually conducive. Their collaborative research throughout the following twenty years at Heidelberg shows how right he was with this assessment. Even the teaching profited from it. There were suddenly more than twenty new enrollments each year in chemistry, for instance. One year, they even exceeded fifty. The newly built chemistry laboratory was decisive in improving the university's attractiveness. Kirchhoff persisted with the endeavors to expand the physics institute to incorporate pure research, against resistance by

---

[12]On Bunsen's vita see, e.g., Georg Lockemann: *R. W. Bunsen. Lebensbild eines deutschen Naturforschers*, Stuttgart: Wiss. Verlagsges., 1949; cf. Bunsen's collected works: *Gesammelte Abhandlungen*, Wilhelm Ostwald and Max Bodenstein (eds.), Leipzig: Engelmann, 1904, 3 vols.; esp. vol. 3 on spectroscopy.

[13]See Peter Borscheid: *Naturwissenschaft, Staat und Industrie in Baden (1848–1914)*, Stuttgart: Klett, 1976; Frank James: Science as cultural ornament, *Ambix* **42** (1995): 1–9; and Jungnickel & McCormmach 1986: 288–290.

the Baden Ministry of the Interior. It paid off. The new physics building, on the drawing board since 1859, could finally be occupied in 1863.[14] At last there was enough room for practical laboratory training. After Leo Königsberger (1837–1921) was appointed to the chair for mathematics in 1869, a *mathematisch-physikalisches Seminar* was instituted, in emulation of Neumann's courses at Königsberg. Under the close supervision of Kirchhoff and Königsberger the students were assigned small research tasks every week that often could be developed into full-fledged research articles or dissertations.[15]

It was around this time that a few events changed Kirchhoff's life. In 1868 he suffered serious injury from a fall and his wife Clara, who had borne him two daughters and two sons, died the following year. Kirchhoff remarried three years later. His second wife, Luise Brömmel, was a physician. Every time Kirchhoff declined an appointment to a university elsewhere, his reputation rose a few notches at Heidelberg — along with his salary level.[16] In 1875 he finally accepted a call to Berlin as Germany's first full professor of mathematical physics and full member of the Prussian Academy of Sciences. This had not been the first attempt to lure him to the capital.[17] Although Kirchhoff taught at the Friedrich Wilhelm University, the Prussian Academy of Sciences funded his professorship. He had been its corresponding member since 1861. In the next eleven years his health steadily worsened, eventually compelling him to relinquish his teaching duties. Health reasons had already been one of the prime motivations for abandoning his research at Heidelberg and coming to Berlin. Nonetheless, Ludwig Boltzmann (1844–1906) wondered at Kirchhoff's indomitable cheerfulness despite having to rely on crutches or a wheelchair. Kirchhoff died of vascular congestion in the brain on October 17, 1887.

---

[14]See *Semper Apertus. 600 Jahre Ruprecht-Karls-Universität Heidelberg 1386–1986*, vol. 5: *Die Gebäude der Universität Heidelberg*, Berlin: Springer, 1985, esp. pp. 336ff., 446ff.

[15]Thirteen students enrolled in this seminar combining teaching and research in its first year, 1870. See Jungnickel & McCormmach 1986, vol. 1: 292f. Arthur Schuster: *The Progress of Physics During 33 Years (1875–1908)*, Cambridge Univ. Press, 1911: 13–14 provides an eye-witness account from one participant.

[16]Jungnickel & McCormmach 1986, vol. 1, p. 291 reports a beginning salary of 2,400 florins that was eventually raised to 6,000 florins a year.

[17]Wilhelm Borchardt: 'Wahlvorschlag fr Gustav Robert Kirchhoff' (1870), in: *Physiker über Physiker*, Christa Kirsten & H.G. Körber (eds.), Berlin: Akademie-Verlag, 1975: 75–79.

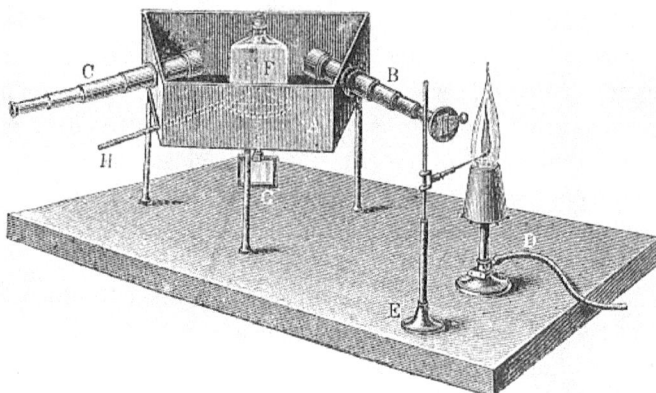

Figure 2: Bunsen's and Kirchhoff's first spectroscope 1859, from Kirchhoff & Bunsen 1860, pl. VI, fig. 1.

Spectrum Analysis

The discovery of spectrum analysis occurred during his Heidelberg period. This result of the close collaboration between Kirchhoff and Bunsen led to their being awarded the first Davy Medal in 1877. Bunsen had been trying to identify the chemical compositions of various salts from their characteristic flame colors since the mid-1850s. He vaporized small samples in a gas burner flame that he and Henry Enfield Roscoe (1833–1915) had optimized for photochemical analysis. Using this burner Bunsen observed the flame colors through color filters in hope of being able to distinguish between often very similarly colored emissions of some elements in the vaporized state. But his results were far from satisfactory. Kirchhoff suggested he work with prisms to decompose the observed light. So he changed over to using a spectroscope, an instrument containing a prism or even a whole series of prisms to break down the rays into their spectral components.

Kirchhoff's unusual expertise in inorganic analysis proved extremely useful here. He soon found in the spectra of a large number of chemical elements what appeared to be characteristic emission lines. They were only observable when the element in question or one of its compounds was held in the flame. The first joint publication by Kirchhoff and Bunsen, titled Chemical Analysis by Spectrum Observations, included a color plate depicting these characteristic spectrum lines for a number of the chemical

Figure 3: Photograph of Gustav Robert Kirchhoff, Bunsen and Henry Enfield Roscoe, taken 1862 in Manchester in the Emery Walker Photostudio; from *The Life and Experiences of Sir Henry Enfield Roscoe*, London & New York: Macmillan, 1906, between pages 72 and 73.

elements.[18] This new method was so sensitive that even trace amounts, say 0.0000003 mg of sodium, still clearly exhibited the characteristic yellow pair of lines the moment it was introduced into the burner flame. By this method Bunsen discovered two new elements in 1860 and 1861. A concentrate from some 44,000 liters of Dürkheimer mineral water revealed strong blue emission lines that Bunsen attributed to what he called caesium (from *caesius*, Latin for sky blue). Rubidium (from *rubidus*, dark red) was isolated

---

[18]Kirchhoff & Bunsen: Chemische Analyse durch Spektralbeobachtungen, *Annalen der Physik* (2) **110** (1860): 161–189 & pl. V–VI and **113** (1861): 337–381. Engl. transl.: Chemical analysis by spectrum observations *Philosophical Magazine* 4th ser. **20** (1860): 89–109, **22** (1861): 329–349, 355–380, 498–510 & pl. V.

out of about 150 kg of lepidolith, a mineral found in Saxony that exhibited prominent red spectrum lines.[19] Although some other elements were discovered by this method (see the subsection below on Kirchhoff's influence), we should not forget that there were numerous false alarms about discoveries of new elements by other experimenters.[20] The success of the spectroscopic method required a rare combination of the perspicuity of a versed spectroscopist as well as the superior analytic knowledge of a chemist. This prerequisite the two Heidelberg professors were able to meet. The latter factor came into play in the provision of pure enough samples and the former, recognizing from among the thicket of spectrum lines of a given specimen the characteristic ones of known elements. Bunsen's and Kirchhoff's published plate did not depict all the lines for these elements. So it was not an easy task. An apparently new line could only be considered a strong indication for an as yet unknown element if it appeared under the normal excitation conditions of the Bunsen-burner flame, without other lines characteristic of known elements appearing.

Emission and Absorption Lines and Kirchhoff's Solar Model

The discovery of these unknown elements bestowed on spectrum analysis canonical status in chemistry. More insights could be gained from these spectroscopic analyses, however. When vaporizing the chemicals in a Bunsen-burner flame, Bunsen and Kirchhoff observed bright lines on a dark background. How did these lines relate to the ones that William Hyde Wollaston (1766–1828) had first observed in 1802? Josef Fraunhofer (1787–1826) had subsequently systematically examined those dark lines in an otherwise continuous colorful solar spectrum in 1814. Almost half a century later, in 1859, Kirchhoff reported:

> I generated a solar spectrum and let the sun's rays pass through a powerful table-salt flame before it fell on the slit. If the solar light was sufficiently subdued, instead of the two dark D-lines, two bright lines appeared; if the intensity exceeded a certain limit, the two dark D-lines presented

---

[19]See Robert Wilhelm Bunsen: Ueber ein neues, dem Kalium nahestehendes Metall, *Monatsberichte der Königlich-Preussischen Akademie der Wissenschaften, Berlin*, 10 May 1860: 221–223; Ueber ein fünftes der Alkaliegruppe angehörendes Element, ibid. (1861): 273–275. Trans.: On a fifth element belonging to the alkali group, *Chemical News* (1861) **3**, no. 80, p. 357.

[20]See V. Karpenko: The discovery of supposed new elements: two centuries of errors, *Ambix* **27** (1980): 77–102.

themselves in much greater sharpness than without the presence of the table-salt flame.[21]

This key experiment revealed the convertibility of the yellow lines of glowing sodium vapor into dark lines (in this case Fraunhofer's D-line). Kirchhoff interpreted the dark spectrum lines as absorption lines and the bright spectrum lines as emission lines.

The exact coincidence in the positions of the dark and bright lines was soon confirmed for a rapidly growing number of elements. Over sixty lines of the iron spectrum were promptly verified, for instance. The probability that these coincidences were governed purely by chance Kirchhoff calculated as one in a trillion! The appearance of dark lines in the sun's spectrum he interpreted as a continuous spectrum generated by a glowing solid core. The gases in the sun's outer layers were absorbing the lines that they for their part emitted. This was the first useful explanation of how the hitherto completely mysterious Fraunhofer spectrum is formed. At the same time it offered a new model of the sun. How the sun's core could remain solid and glow despite such evidently extreme temperatures remained puzzling. The discovery that under suitable conditions even gases can emit continuous spectra later invalidated Kirchhoff's model. Nevertheless, the prospect of being able to determine the chemical composition of the sun from the presence or absence of certain spectrum lines motivated Kirchhoff to compile a detailed chart of the solar spectrum.[22]

The ensuing priority dispute about Kirchhoff's and Bunsen's discoveries revealed that George Gabriel Stokes (1819–1903) had already mentioned this observation in his lectures ten years earlier. He had specifically told William Thomson (1824–1907) about the coincidence of the sodium lines in the yellow region with the bright lines of the sodium flame, interpreting it as a resonance effect. Jonas Anders Ångström (1814–1874) had also noticed it while working on experiments seeking to explain spectrum lines around 1855. All the same, consensus was soon reached that Kirchhoff had been the first to fully understand that each substance absorbs precisely the same spectrum lines that it emits.

---

[21] Kirchhoff: Ueber die Fraunhofer'schen Linien, *Monatsberichte der Akademie zu Berlin* (1859): 662–665, esp. p. 663. Our translation. It appeared in English in the following year: Fraunhofer's lines, *Philosophical Magazine* (4th ser.) **19** (1860): 193–197.
[22] For a contextualization and detailed analysis see Klaus Hentschel: *Mapping the Spectrum. Techniques of Visual Representation in Research and Teaching*, Oxford Univ. Press, 2002, esp. pp. 124ff.

Kirchhoff's Law and the Kirchhoff Function

In 1860 Kirchhoff also developed more general ideas about the reversability of spectrum lines and the close relationship between radiative emission and absorption. They culminated in what is called 'Kirchhoff's law.' Emissive and absorbing powers are at a constant ratio $J[I, \lambda]$ for a given wavelength λ at temperature T, independent of the material properties of the examined body.

> The quantity denoted as J is [...] a function of the wavelength and the temperature. It is a task of high importance to find this function. Great problems stand in the way of its experimental determination; nevertheless the hope of being able to find it out by experiment seems well-founded, because it is undoubtedly simple in form, as are all functions we have hitherto become acquainted with that are not dependent on the properties of individual bodies.[23]

Determining this function J by experiment or even theoretically was certainly not a trivial matter. It was eventually achieved toward the end of that century using what is known as a black body, an object that completely absorbs any radiation falling on it. The way Kirchhoff proved his law was brilliant.[24] He used considerations about the exchange of radiation between bodies whose emissive and absorbing behaviors are clearly defined by means of thermodynamic equilibrium conditions. No atomistic or material properties were needed at all. This method later served as a model for Planck and Einstein.

Kirchhoff's Style and Ideal Method

Other articles by Kirchhoff span a wide range of subjects in the areas of induced magnetism, electrostatics, hydrodynamics, thermal conduction,

---

[23] Kirchhoff: Ueber das Verhältnis zwischen dem Emissionsvermögen und dem Absorptionsvermögen der Körper für Wärme und Licht, *Annalen der Physik* (2) **109** (1860): 275–301, esp. p. 292. Our translation. The paper also appeared in English transl.: On the relation between the radiating and absorbing powers of different bodies for light and heat, *Philosophical Magazine* (4th ser.) **20** (1860). See also Daniel M. Siegel: Balfour Stewart and Gustav Robert Kirchhoff: Two independent approaches to Kirchhoff's radiation law, *Isis* **67** (1976): 565–600.

[24] For details cf. also Arne Schirrmacher: Experimenting theory: The proofs of Kirchhoff's radiation law before and after Planck, *Historical Studies in the Physical and Biological Sciences* **33**, (2003): 299–335.

conveyance of sound in narrow tubes, and elasticity theory. In an honorary address in 1887 Boltzmann — true to the nationalistic worldview of his day — stylized the descriptive approach Kirchhoff used in his papers and lectures as "a prototype of the German way of treating problems in mathematical physics." It certainly starkly contrasted the complex argumentation of Maxwell's theory of gases. Boltzmann specified the characteristics of this "German way" as: "sharpest precision in the hypotheses, fine-grain polishing, steady, rather epic progress with steely consistency, no hushing up of any problems, illuminating the slightest shadow of a doubt."[25] Boltzmann was clearly enthralled.

> Particularly among [...] Kirchhoff's articles, some are of uncommon beauty. Beauty? I hear you ask; do the Graces not flee where integrals crane their necks? Can anything be beautiful for which the author lacks the time to supply even the tiniest of superficial ornaments?– Indeed.– Precisely this simplicity, this indispensability of every word, every letter, every hyphen, brings the mathematician closer than all other artists to the Almighty; unequalled among the other Arts, it establishes the sublime — resembling, at best, only symphonic music.[26]

To Kirchhoff mechanics posed as the ideal method for doing science.[27] He compared it to geometry because both these sciences were "applications of pure mathematics" whose laws stood, "with regard to their certainty, on exactly the same rung." For, "absolute certainty may be ascribed to mechanical laws with the same right as to geometrical ones." This excerpt from Kirchhoff's vice-rectorial speech from 1865 shows how firmly rooted he still was in the deterministic view of the world comprehending classical physics. Not long afterwards this view would be shaken in its foundations by Boltzmann, Planck and Einstein, building upon his work. Kirchhoff confidently asserted:

> If all forces of Nature were known, and if one knew in what state matter is at *one* point in time, then it would be possible to figure out its state for any later point in time by mechanics and derive how the manifold phenomena of Nature follow and attend one another. The highest goal that the Sciences have to seek is [...] tracing all natural phenomena back to mechanics. This

---

[25]See Ludwig Boltzmann: G.R. Kirchhoff' (Festrede 1888), in: *Populäre Schriften*, Leipzig: Barth, 1905, p. 74.

[26]Boltzmann 1905, pp. 73f.

[27]See, e.g., R. Helmholtz: Gustav Robert Kirchhoff (Nachruf), *Deutsche Rundschau*, issue 14, vol. 54 (1888): 233–245, quote from p. 243; Jungnickel & McCormmach [1986] vol. 1: 303–304, vol. 2: 125–128.

goal of Science will never be completely reached; but the mere fact that it is recognized as such affords a certain satisfaction and nearing it is one of the highest pleasures that employment with Nature's phenomena may grant.[28]

Mechanics was not just supposed to serve as a model for the developing physical disciplines in the nineteenth century, such as heat theory, electrodynamics and optics. It was also supposed to be the methodological ideal of a strictly phenomenalistic science. This is particularly evident in Kirchhoff's introductory words to his *Mechanics*, a work based on lectures he held at Heidelberg shortly before moving away to Berlin. Kirchhoff saw the purpose of mechanics not as 'explanation' from real causes but as the "complete and simplest of descriptions" by means of mathematics. His mechanics lecture introduces the concepts of force and mass as helpful or "auxiliary concepts" (*Hülfsbegriffe*). Force is defined by the acceleration a material particle experiences per unit time. In principle, knowledge about all these 'accelerating forces' would suffice for a description of the world. Convenience alone has these accelerations multiplied by "a certain positive constant; this constant is called the mass of the moving particle."[29] Kirchhoff also tried to reduce the theory of heat to the basic building blocks of 'space,' 'time' and 'matter.' Matter was preferentially treated in conformity with the notion of a continuum. Only after the definition for 'heat' referred to material motion did Kirchhoff discuss the atomic model of matter and the first steps Clausius and others were taking at the time toward a kinetic theory of heat. As it turned out, this attempt at a 'metaphysics-free' abstention from hypotheses was not the way to find the simplest possible description of natural phenomena. Even so, Kirchhoff's approach exerted a strong influence on his contemporaries. Particularly Heinrich Hertz, Ernst Mach and Wilhelm Ostwald, as well as a few twentieth-century philosophical interpreters of quantum mechanics, chose to view it merely as an instrument for describing and predicting detectible quantities (observables). But it was not universally welcomed. The Göttingen mathematician Felix Klein (1849–1925) wrote in his lectures on the development of nineteenth-century mathematics:

> The style emanating from Kirchhoff's writings is essentially the dominant one in the mathematical physics of some decades: the highest law was regarded as avoidance of premature hypotheses, if not errors, and repression

---

[28]Kirchhoff: *Über das Ziel der Naturwissenschaften*, Heidelberg: Mohr 1865.
[29]See Kirchhoff: *Vorlesungen*, vol. 1: *Mechanik*, p. 23.

of any emotional engagement, joy of discovery or amazed wonderment before the boundless mystery of the world of phenomena. [...] Completely directed toward the management of what is available, Kirchhoff seems to have found new discoveries inconvenient or of lesser interest. [...] I cannot conceal that I find this notion of science extremely unappealing, because it subdues the joy of learning and the drive for the pursuit of research.[30]

Influence

Considering the enormous importance of Kirchhoff's circuitry laws for modern electrical engineering, it is astonishing how long it took for their practical relevance to be recognized in France. Two decades passed before the telegraph technician Jules Raynaud (1843–1888) became acquainted with them.[31] This despite the fact that an underwater cable had been linking Calais to Dover since 1851 and the telegraph installations throughout Europe were rapidly becoming an increasingly dense network.

Quite the opposite for the reception of the new discipline of spectroscopy, triggered by Kirchhoff's and Bunsen's joint research.[32] Others quickly snatched up Bunsen's recipe for discovering elements; and numerous newcomers soon appeared on the table of the chemical elements.[33] Systematic comparisons between the terrestrial emission spectra of the

---

[30] Felix Klein: *Vorlesungen über die Entwicklung der Mathematik im 19. Jahrhundert*, Berlin: Springer, 1926, vol. 1, pp. 219f.

[31] Textbooks for the elite *École Polytechnique* for engineers started discussing Kirchhoff's laws around 1870; that was when they were also being routinely applied in installations of the nation's growing cable network. Andrew J. Butrica: The 'rediscovery' of Kirchhoff's circuitry laws, in: *Beyond History of Science*, Elisabeth Garber (ed.), Bethlehem: Lehigh Univ. Press, 1990, pp. 208ff.

[32] See William McGucken: *Nineteenth-Century Spectroscopy. Development of the Understanding of Spectra*, Baltimore: Johns Hopkins Univ. Press, 1969. Heinrich Schellen: *Die Spectralanalyse in ihren Anwendungen*, Braunschweig: Westermann, 1870, and Hentschel 2002 (note 22).

[33] Caesium and rubidium by Bunsen 1860–61 (cf. Bunsen: Rubidium and caesium, *Chemical News* (1861) **3**, no. 80, pp. 44, 286), thallium by William Crookes 1861 (1832–1919), indium by Ferdinand Reich 1863 from the Mining Academy in Freiberg, gallium by Paul Émile François Lecoq de Boisbaudran 1874 (1838–1912), scandium by Nilson 1879, germanium by Winkler 1886, and finally helium. The French astronomer Janssen and the British astrophysicist Joseph Norman Lockyer (1836–1920) had each independently detected this element in the solar spectrum in 1868. William Ramsey (1852–1916) managed to isolate it out of the mineral cleveite in 1898 and Crookes was thus able to analyse it, demonstrating the wavelength-identity of its spectral lines with the dark lines attributed to helium in the Fraunhofer solar spectrum.

various chemical elements or compounds and the solar spectrum, stellar and nebular spectra yielded important information about their physical and chemical compositions. Slight differences in the appearance and positions of these extraterrestrial spectrum lines gave clues about the physical conditions and radial velocities of otherwise inaccessible objects. Redshifts in stellar spectra, interpreted in the mid-1860s as Doppler shifts, led to the discovery of radial motions of the stars relative to the Earth by William Huggins (1824–1910). In 1908 George Ellery Hale (1868–1938) detected magnetic fields in solar spots using the Zeeman effect, thus opening the way to determining the solar magnetic field spectroscopically. So it was not surprising that Kirchhoff would be offered the directorship of the newly founded Astrophysical Observatory in Potsdam in 1874 — which he declined, however, for age reasons. Under the leadership of Hermann Carl Vogel (1841–1907) its main focus became the identification of Doppler shifts in stellar spectra. The roots of all these astrophysical investigations reach back to Kirchhoff's papers of 1861.[34]

From Kirchhoff to Quantum Theory

The ideas formed as a direct result of Kirchhoff's radiation theory by Wilhelm Wien (1864–1928) and particularly by Max Planck (1858–1947) fed the development of quantum theory around 1900. Planck succeeded Kirchhoff on the chair for theoretical physics in Berlin and also edited the last two volumes of his published lectures.[35] The first attempt to determine experimentally Kirchhoff's function $J[\lambda, T]$, describing the material-independent relation between emissive and absorbing powers, was made by 1894 Friedrich Paschen (1865–1947) in 1894. The glowing body he used to approximate Kirchhoff's 'black body' was still very poor, however. Later investigations had the radiation released from a tiny opening

---

[34] See Kirchhoff: Untersuchungen über das Sonnenspektrum und die Spectren der chemischen Elemente, *Abhandlungen der Akademie der Wissenschaften zu Berlin* (1861): 63–95 & pl. I–III; (1862): 227–239 & pl. I ("containing observations by K. Hoffmann"). See also Kirchhoff: Zur Geschichte der Spectral-Analyse und der Analyse der Sonnenatmosphäre, *Annalen der Physik* (2) **118** (1863): 94–111.

[35] The third and fourth volumes: *Elektrizität und Magnetismus*, published in 1891 and *Theorie der Wärme* in 1894.

in a cavity with walls held at a constant temperature. This experimental set-up, Kirchhoff had suggested in 1860:

> If a space is enclosed by bodies of equal temperature and no rays can penetrate through these bodies, then the quality and quantity of each beam in the interior of the space is composed as if it came from a perfectly black body of the same temperature, thus it is independent of the properties and shape of the bodies and only dependent on the temperature.[36]

Measurements in the infrared were conducted on this artificial black body by Otto Lummer (1860–1925) and Ernst Pringsheim (1859–1917) as well as around 1900 by Heinrich Rubens (1865–1922) and Friedrich Kurlbaum

Figure 4:   Portrait of Gustav Robert Kirchhoff c. 1880, from a photogravure by the *Photographische Gesellschaft Berlin*, no. 8126, copper engraving and print by A. Weger in Leipzig, first published in Kirchhoff's collected works: *Gesammelte Abhandlungen*, 1882. Frontispiece.

---

[36]Kirchhoff 1860, p. 300. Our translation.

(1857–1927) at the German bureau of standards, the *Physikalisch-Technische Reichsanstalt* in Charlottenburg, a suburb of Berlin. Planck analysed their results theoretically and his famous radiation formula was the outcome. His was the first complete and consistent expression for the entire energy spectrum of a black body consistent with all the experimental data from the various energy regions.[37]

Thus Kirchhoff not only worked on topics from all branches of classical physics but also led the path to its twentieth-century nonclassical continuations.

Prof. Klaus Hentschel
Head of the Section for History of Science & Technology
University of Stuttgart
Keplerstr. 17
D-70174 Stuttgart, Germany
e-mail: klaus.hentschel@hi.uni-stuttgart.de
http://www.uni-stuttgart.de/hi/gnt/hentschel

---

[37]On the theoretical contexts see, e.g., Hans-Georg Schöpf: *Von Kirchhoff bis Planck: Theorie der Wärmestrahlung in historisch-kritischer Darstellung*, Braunschweig: Vieweg, 1978, Thomas S. Kuhn: *Black-Body Theory and the Quantum Discontinuity 1894–1912*, Chicago: Univ. of Chicago Press 1988, and Jochen Büttner *et al.*: *Revisiting the Quantum Discontinuity*, Max Planck Institute Preprint no. 150 (2000); on Lummer's experimental work see, e.g., Dieter Hoffmann's contribution to the last-mentioned anthology of papers.

Kirchhoff's Theory of Diffraction as Presented in His 1882 Treatise, Commentary by Ning Yan Zhu

Dedicated to Prof. F.M. Landstorfer,
inventor of the eponymous antenna, on the occasion of his 75th birthday.

Prologue

Ever since its enunciation in 1882 in the Transactions of the Royal Prussian Academy of Sciences [12] and one year later in the more widely read *Annalen der Physik* [13], Kirchhoff's approach (also known as 'Kirchhoff's method', or 'Kirchhoff's approximation') has been — in spite of its heuristic nature — one of the most widely used high-frequency computational methods in wave physics. For a thorough discussion of 'Kirchhoff's approach' and the eponymous theory of diffraction based on it the readers are referred to [1, 8, 27]; see also [17].

In recent years, however, several publications, written by authors working in the field of applied electromagnetics, claim that this very method was not invented by Kirchhoff:

For example, in an otherwise well-written review [4] which cited explicitly both the original paper of Kirchhoff [12, 13] as well as its French translation of 1886 [14], it is stated that "Kirchhoff's approximation is generally but erroneously known as the Physical Optics approximation. Indeed, very often the Kirchhoff's approach is confused with the Physical Optics approach. It should be noted that Kirchhoff's approach is applied to determine the field behind an opaque screen with an aperture, when the field over the aperture is assumed to be unperturbed. ... On the other hand, the Physical Optics approach refers to the field scattered by a reflecting surface, S. The scattered field on the same surface is approximately determined by the reflection coefficients. This was first formulated, in the case of an electromagnetic wave, by H. M. Macdonald. In conclusion, it is

more correct to refer to the Physical Optics approximation as an extension
of Kirchhoff's approach." Clearly, what is termed the Physical Optics
approach in applied electromagnetics refers to an approximate computational method, whereas in optics, physical optics serves as a synonym for
wave-optics [31]. Macdonald's paper of 1912 [18] — that is 30 years after
Kirchhoff's approach was published — employed the geometrical-optical
approximation for the fields on the surface of a scattering body in the
Huygens' principle for electromagnetic fields.

In his recent works, Ufimtsev, one of the leading experts of the so-called
Physical Theory of Diffraction, an extension of Kirchhoff's approach, has
also adopted this point of view (see for example, [35, p. 11]).[1]

A fair appreciation of Kirchhoff's approach must recall the content of
the original paper of Kirchhoff. To this end, the remainder of the present
article gives a brief account of the contributions made by Kirchhoff in his
1882 paper. An English translation of Kirchhoff's paper is included as the
next article in this book.

For the sake of convenience, the 1882 paper by Kirchhoff will be denoted
below as 'the Paper,' and the numbered equations refer to those of the
Paper.

The Paper

The 29 pages of the paper of particular interest to us here contain
Kirchhoff's contributions to diffraction theory. The essential parts of the
material were the subject of his "Lectures on Mathematical Optics" already
taught four times (every second winter semester from 1875 through 1881) at
the University of Berlin [9, pp. 281–282]. After Kirchhoff's untimely death
in 1887, a textbook based on his lecture notes appeared under the editor-
ship of a mathematician, Kurt Hensel, as the second volume of Kirchhoff's
"Lectures on Mathematical Physics" [15]. The material contained in the
Paper was expounded upon in Lectures 1, 2, 3, 5, and 6 of [15].

As is characteristic for the 19th century, Kirchhoff's study is based on
the elastic aether theory. Interestingly, the laws of wave-motion in such an

---

[1]In the second edition of Ufimtsev's monograph published in July 2014, it reads now
[36, p. 7]: 'In electromagnetics, the term *physical optics* (PO) is usually applied to the
high-frequency asymptotic technique introduced by Kirchhoff (1883) and Macdonald
(1912).' 'Kirchhoff (1883)' refers to the reprint of Kirchhoff's paper by *Annalen der
Physik* [13] which is cited by Ufimtsev [36, p. 402].

elastic medium liken those by Maxwell ([16, pp. 7–8], [3, pp. xxix–xxx], [5], [6]). Even though the elastic theory of light has been abandoned in favor of Maxwell's electromagnetic theory, this does not at all mean that earlier techniques like those of Kirchhoff are now obsolete.

Mathematical Formulation of Huygens' Principle

In Kirchhoff's opinion, the arguments used at that time to explain the formation of light rays, their reflection and refraction as well as the phenomena of diffraction, lack rigor in many aspects, but it is possible to sharpen these arguments. To this end, the Paper derives in §2 a theorem which will make Huygens' principle more precise and more general.

To derive the theorem, the Green's formula (the first equation of §2) is applied to two functions $\mathfrak{U}$ and $\mathfrak{V}$ inside a bounded volume which is free of luminous points. Since these two functions satisfy the wave-equation (1), one side of Green's formula can be written as a partial derivative of a volume-integral with respect to time t. Integrating both sides of the resulting equation in a suitable time-interval $(-t', t'')$, one arrives at equation (5).

Then an impulse point-source, that is, a point-source whose dependence upon time t is given by what is known today as the Dirac delta-function $F(\zeta)$, is taken for $\mathfrak{V}$. After a suitable choice of t' and t'', terms related to the volume-integral disappear. In order to enable the application of Green's formula for a point o inside the bounded volume, a sphere of a small radius centered at o is excluded, leading to equation (7).

As the radius of the small sphere tends to zero, the related surface-integral yields merely one term. Accounting for the property of $F(\zeta)$ yields the formula (10), an integral representation for $\phi_o(t)$, the value of the wave function ϕ at the point o and time instant t, in terms of the values of ϕ and of its normal derivative at earlier time instants on the closed surface of the bounded volume.

In the next step Kirchhoff shows that the same formula (10) is also valid when all the luminous points are inside the closed surface s while the point o is outside of it. Lastly, he deduces the formula (12) which describes ϕ_o, the total field at point o, as a linear superposition of ϕ_o^*, the incident field at that point due to a luminous point at point ι, and $\int \Omega ds$, the contribution of equivalent sources induced on the surface of a foreign body by the point-source.

Formula (12) represents the first mathematical formulation of Huygens' principle for arbitrarily time-dependent wave-motion. Hence it is also termed the Huygens-Kirchhoff principle [16, 27].

Formula (12) extends the result by Helmholtz, which holds for time-harmonic wave-motion [7]. Considering that the light source is turned on a finite time instant ago, no radiation conditions have to be imposed on the wave field in the derivation of (12), as is necessary in case of Huygens' principle for time-harmonic waves in an unbounded space. Besides, Sommerfeld formulated radiation conditions in 1912 [30].

According to ([33, p. 176], [24, p. 377]), Kirchhoff pioneered the use of what is now known as the Dirac delta-function in physics; see his 1876 textbook on mechanics, the first volume of his "Lectures on Mathematical Physics" [11, p. 315].

Today's derivation of Huygens' principle basically follows Kirchhoff's paper of 1882, see for instance [10, Sec. 1.17].

Method of Stationary Phase for Double Integrals

Beginning at §3, the Paper is centered on time-harmonic light waves.

In order to draw conclusions from the rigorous formulation of Huygens' principle (12), one needs a technique for evaluating the surface integrals in the optical regime with infinitesimal wavelength λ.

When the Paper appeared, the principle of stationary phase was barely known [38, pp. 229–230]. To apply this principle to a highly-oscillating double integral as contained in Huygens' principle (12) requires a detailed analysis. §3, the longest in the Paper, serves precisely this purpose.

To simplify the analysis, the discussion in §3 is confined to field points outside of the caustics of the surface and of its rim, and well away from the shadow boundaries. As pointed out in [37], the key step lies in reducing the double integral to a one-dimensional one by using the concept of Fresnel zones.

It is found that the isolated critical points of the first kind [2, 37, 39] of the phase, here proportional to the total distance from the source point ι via a point on the surface s to the field point o, correspond to reflection points or intersecting points of the line connecting points ι and o with the surface s.

For convenience, a new Cartesian co-ordinate system is introduced with its origin coinciding with such a point on surface s, with the surface normal as the z axis. In such a system, the neighborhood of surface s can be approximated in terms of (21).

Expanding the phase in powers of x and y and dropping terms higher than second order, the asymptotic value of the integral under study is given in formulae (24, 25), if the Fresnel zone is of elliptic shape. If the Fresnel zone is of hyperbolic shape, then the double integral is asymptotically equal to (26, 27).

The results of the method of stationary phase are applied to the case when the line connecting points ι and o meets surface s. In this way it is shown from Huygens' principle that the secondary sources excited by a spherical wave (3) again produce a spherical wave (4).

Today's treatment of the method of stationary phase for double integrals resembles the procedure by Kirchhoff, with the method of resolving multiple integrals used to effect the reduction of the surface integral to a one-dimensional one [39].

Kirchhoff's Approach

What remains to be done is to determine the values of ϕ and its normal derivative $\dfrac{\partial \phi}{\partial N}$ on the surface of the foreign body (§4). In what follows, Kirchhoff considers time-harmonic light sources.

Kirchhoff recalls at first the well-known relationships between the incident and reflected plane-waves at a planar interface, namely equations (30) where, in today's language, $c \exp(-i\gamma)$ denotes the complex reflection coefficient for the wave function ϕ, provided a time-dependence $\exp(-i2\pi t/T)$ is assumed.

Then Kirchhoff continues (on original p. 660): "These laws can be applied to the case to which equation (12) refers, when the wavelength λ is presumed to be infinitesimal and the curvature of the surface of the imagined body is nowhere supposed to become infinitely large."

Under these conditions, a close proximity of each point to the smooth surface of the imagined body can be replaced by the tangential plane at that point, and the reflected or transmitted waves are determined in terms of equations (30). Expressed in today's terminology, this means nothing else than to use the geometrical-optical approximation for the fields on the surface of the imagined body.

Replacing ϕ and its normal derivative on the surface of an imagined body with their respective geometrical-optical approximations is usually called 'Kirchhoff's *Ansatz*,' whereas using Kirchhoff's Ansatz in Huygens' principle is regarded as 'Kirchhoff's approach.' The surface integrals contained in (12) are usually evaluated numerically. As is well known, Kirchhoff's approach is the basis of his theory of diffraction.

The conjectured asymptotic nature of Kirchhoff's Ansatz has been proven, to the author's knowledge, only for such simple cases as a strictly convex, smoothly bounded and acoustically soft body illuminated by a plane wave. In this case Kirchhoff's Ansatz represents the first term of an asymptotic series; in a neighborhood of the shadow boundaries, the second term of the same series is, in comparison with the first one, no longer negligible [32, 34].

Equipped with this Ansatz and the method of stationary phase for double integrals, Kirchhoff is now ready to draw conclusions from Huygens' principle (12) and outline his theory of diffraction.

Scattering by a Black Body

The very first example deals with the influence of a black body upon light emitted by a point-source (bottom of original p. 660). A black body is a body that neither reflects incident waves nor lets them pass through it. A cone whose apex lies in the luminous point and whose surface touches the black body divides the surface of the latter into two parts. According to Kirchhoff's Ansatz, the total field and its normal derivative are identical to those of the incident wave on the lit part, and disappear on the shady part.

Using these values in equation (12) and employing the method of stationary phase derived in §3, it is shown that the field value at a point o is given by the incident wave alone, as long as the line connecting the source point with the field point does not intersect the black body; otherwise the field vanishes completely. Therefore, a black body casts a shadow and the light from a luminous point propagates in a rectilinear manner as independent rays.

Light Transmission through an Aperture

If the conditions for deriving the results of the method of stationary phase in §3 are violated, then the conclusions drawn are no longer valid. In this case the phenomena of diffraction prevail.

To study these phenomena, Kirchhoff considers a luminous point enclosed by a black screen with an aperture (§5). Now the closed surface is made up of the invisible part of the black screen and the aperture surface, say. To determine the field strength outside of such a closed surface, equation (10) is needed. According to Kirchhoff's Ansatz, the surface integration needs only to be extended over the aperture, and the field strength there is replaced by that of the incident wave.

Kirchhoff points out that phenomena of diffraction appear for a field point when the distance from the source, via either a finite part of the aperture or a finite part on the rim of the aperture to this point, is stationary, and for a point close to the shadow boundary.

Then Kirchhoff calculates the diffraction of a luminous point by an aperture. To simplify the matter, he assumes that the aperture is planar and its dimensions are sufficiently small in comparison with both the distance from the aperture to the source ι and the distance from the aperture to the field point o. The formulation of the problem also includes the diffraction of a luminous point by a circular aperture with the source on the axis of the aperture. Fresnel observed this kind of diffraction phenomena (cf. original p. 663). Kirchhoff presents a closed-form formula which agrees well with numerous experimental results published in the *Annalen der Physik*.

Later, Maggi [20] and in particular Rubinowicz [25] showed that the result obtained from Kirchhoff's approach applied to light transmission through an aperture can be split into two parts in a rigorous manner: a ray-optical part and a line integral along the rim of the aperture that describes the edge diffraction. Applying the method of stationary phase to the line integral, Rubinowicz discovered in 1924 what is now called the cone of diffraction [26]. Actually an earlier observation of such a cone of diffraction was made by Maey in 1893. In his Königsberg dissertation he applied Kirchhoff's approach to diffraction of light waves by a half-plane; he found the cone of diffraction in this special case and confirmed it experimentally [19].

It is worth mentioning that around 14 years after the Paper, Arnold Sommerfeld, another great son of Königsberg, succeeded in obtaining an exact closed form solution to diffraction by a perfectly conducting half-plane [28].

Light Transmission through a Grating

The geometrical dimension of the aperture considered in the preceding case is much larger than the wavelength, hence justifying the use of Kirchhoff's Ansatz. In the case of a grating (§6), the width of the engraved grooves is usually of a few wavelengths, therefore, Kirchhoff's approach cannot be expected to furnish reasonable results in this case. This notwithstanding, the positions of the maximum intensities of light are accurately predicted by Kirchhoff's approach, as experiments have confirmed.

Kirchhoff offers an explanation by considering the scattering of normally incident light on a planar grating of rectangular size ($2b \times 2ne$), where $2b$ denotes the length of the grooves, e the period of the grooves and $2n$ the number of the grooves. For a field point in the far-field region, the integrals contained in Kirchhoff's approach can be given in closed form. It follows from these formulae that the light intensity becomes infinitely large at the same angles, as the maxima of the light intensity known from experiments.

Reflection and Transmission of Light at the Surface of a Body

In the last section (§7), Kirchhoff considers an arbitrarily shaped body illuminated by a point source. To simplify the matter, Kirchhoff completely covers the body with a black layer excepting a small area on the lit side of the body. In line with Kirchhoff's Ansatz, the values of ϕ and its normal derivative are given in the following way: on the free area of the lit part, they are given by a linear superposition of the incident and reflected rays with appropriate reflection coefficient; on the remaining lit part covered by the black layer, they are given by the incident ray alone; on the shady side, they disappear.

Applying the method of stationary phase, it is shown that a reflected field exists at a point outside of the body if and only if for such a point there exists a point on the free area whose connecting lines with the source and the field-points subtend the same angle with the surface normal at that point.

In addition, the method of stationary phase reveals the geometrical properties of a bundle of rays which originate from the luminous point and are reflected off a curved surface, and shows how the intensity and phase of such a ray-bundle vary from one point to another.

At the end of the Paper, it is pointed out that a similar study of refracted rays can likewise be carried out, as detailed in Lecture 4 of [15].

Hence Kirchhoff's Paper establishes for the very first time a connection between wave-optics and ray-optics by the approach rightly named after him. As is well known, an alternative way of demonstrating such a connection, departing from the Helmholtz equation, was described almost 30 years later by Arnold Sommerfeld and Iris Runge, based on an idea by Debye [29].

The Vectorial Nature of Kirchhoff's Theory of Light

Kirchhoff was well aware of the fact that the so-called displacement vector $\vec{B} = (u, v, w)$ must be divergence-free (the incompressibility condition), that is, in today's vector notation,

$$\operatorname{div} \vec{B} = 0.$$

And in order to meet this requirement, the simplest permissible way to construct the displacement is by means of the so-called wave functions $\vec{A} = (U, V, W)$ according to

$$\vec{B} = \operatorname{rot} \vec{A}.$$

Therefore, Kirchhoff obtained a divergence-free displacement by representing it by the wave functions which are then of scalar nature. This fact was well known at that time (see for instance [20]) and was emphasized by Maey in his later study of light diffraction by a half-plane [19].

This fact seems to have been thoroughly forgotten in the meantime. 120 years later, two papers appeared [22, 23] which, after first criticizing the alleged scalar nature of Kirchhoff's approach, proposed a way to render it applicable to electromagnetic fields. Their way consists in applying the "scalar" approach by Kirchhoff to the components of the Hertzian vector $\vec{\Gamma}$. Because the Hertzian vector $\vec{\Gamma}$ is related to the vector potential $\vec{A}$ via

$$\vec{A} = \mu\epsilon \frac{\partial \vec{\Gamma}}{\partial t},$$

their approach practically replicates that of Kirchhoff.

Epilogue

As outlined in the preceding section, Kirchhoff's 1882 paper, a synopsis of his lectures read at the University of Berlin, contains a wealth of seminal contributions to mathematical physics in general and to the theory of diffraction in particular. It is written in a surprisingly modern language, except that nowadays Maxwell's equations are the departing point for such a study and the method of stationary phase for double integrals is part of the toolkit of every serious student of diffraction theory.

After having clarified what Kirchhoff's approach is as formulated in his 1882 paper, we can turn our attention to its relationship with the Physical Optics approximation. Using the usual definition, such as the one in [4], the Physical Optics approximation is one part of Kirchhoff's approach. As rightly pointed out in [4], the so-called aperture method used in calculating the radiation properties of large reflector antennas represents a special case of Kirchhoff's approach.

As expounded at the end of the preceding section, an electromagnetic wave can be correctly described in terms of Kirchhoff's approach when one follows the rules set by Kirchhoff, that is, to express each component of the vector potential $\vec{A}$ in terms of (12), rather than each component of the electric or magnetic field.

In the preface to his *Treatise*, Maxwell wrote [21, p. xi]: "It is of great advantage to the student of any subject to read the original memoirs on that subject, for science is always most completely assimilated when it is in the nascent state, . . . ".

The 1882 paper by Kirchhoff doubtlessly belongs among these classics well worth being read, now as much as it has been in the past. Exactly for this reason it is high time to make it accessible to English readers, 134 years after its original publication.

References

[1] Baker, B. B. & E. T. Copson: *The Mathematical Theory of Huygens' Principle*, 3rd ed., New York: Chelsea Publishing Company, 1987.

[2] Bleistein, N. & R. A. Handelsman: *Asymptotic Expansions of Integrals*, New York: Dover Publications, 1986.

[3] Born, Max & Emil Wolf: *Principles of Optics*, 7th (exp.) ed., Cambridge: Cambridge University Press, 2003.

[4] Bucci, O. M. & G. Pelosi: From wave theory to ray optics, *IEEE Antenna Propagat. Magazine* 36(4) 1994: 35–42.

[5] Darrigol, Olivier: *Electrodynamics from Ampére to Einstein*, Oxford: Oxford University Press, 2000.

[6] Darrigol, O.: *A History of Optics*, Oxford: Oxford University Press, 2012.

[7] Helmholtz, Hermann v.: Theorie der Luftschwingungen in Röhren mit offenen Enden, *Journal für die reine und angewandte Mathematik* 57(1) (1860): 1–73.

[8] Hönl, Helmut, A. W. Maue & K. Westpfahl: *Theorie der Beugung*, in *Handbuch der Physik*, vol. XXV/1, ed. by Siegfried Flügge, Berlin: Springer, 1961.

[9] Hübner, Kurt: *Gustav Robert Kirchhoff*, Ubstadt-Weiher: Verlag Regionalkultur, 2010.

[10] Jones, D. S.: *Acoustic and Electromagnetic Waves*, Oxford: Clarendon Press, 1986.

[11] Kirchhoff, Gustav Robert: *Vorlesungen über Mathematische Physik*, vol. 1: *Mechanik*, Leipzig: Teubner, 1876; online available at http://archive.org/stream/vorlesungenberm02kircgoog#page/n7/mode/1up

[12] Kirchhoff, G.R.: Zur Theorie der Lichtstrahlen, *Sitzungsberichte der Königlich Preußischen Akademie der Wissenschaften zu Berlin* 15 (1882): 641–669.

[13] Kirchhoff, G.R.: Zur Theorie der Lichtstrahlen, *Annalen der Physik* (3rd ser.) 18(254) (1883): 663–695.

[14] Kirchhoff, G.R.: Sur la théorie des rayons lumineux, *Annales Scientifiques de l'École Normale Supérieure* (3rd ser.) 3 (1886): 303–342.

[15] Kirchhoff, G.R.: *Vorlesungen über Mathematische Physik*, vol. 2: *Mathematische Optik*, Leipzig: Teubner, 1891. Available online at http://archive.org/stream/vorlesungenberm01plangoog#page/n14/mode/1up

[16] Kline, M., & I. W. Kay: *Electromagnetic Theory and Geometrical Optics*, New York: Interscience Publishers, 1965.

[17] Kravtsov, Yu. A. & N. Y. Zhu: *Theory of Diffraction. Heuristic Approaches*, Oxford: Alpha Science, 2010.

[18] Macdonald, H. M.: The effect produced by an obstacle on a train of electric waves, *Phil. Trans. Roy. Soc. Lond. A* 212 (1912/13): 299–337.

[19] Maey, E.: Ueber die Beugung des Lichtes an einem geraden, scharfen Schirmrande, *Annalen der Physik* (3rd ser.) 49(5) (1893): 69–104.

[20] Maggi, G. A.: Sulla propagazione libera e perturbata delle onde luminose in un mezzo isotropo, *Annali di Matematica pura ed applicata* (II) 16 (1888): 21–48.

[21] Maxwell, James Clerk: *A Treatise on Electricity & Magnetism*, Vol. 1, unabr. 3rd ed., New York: Dover Publications, 1954.

[22] Nesterov, A.V. & V.G. Niziev: Vector solution of the diffraction task using the Hertz vector, *Physical Review* E 71 (2005): 046608.

[23] Niz'ev, V.G.: Dipole-wave theory of electromagnetic diffraction, *Physics Uspekhi* 45(5) (2002): 553–559.

[24] Rosas-Ortiz, O.: On the Dirac-Infeld-Plebański delta function, in *Topics in Mathematical Physics, General Relativity and Cosmology in Honor of Jerzy Plebański*, ed. by H. García-Compeán *et al.*, Singapore: World Scientific, 2006, pp. 373–385.

[25] Rubinowicz, Adalbert Wojciech: Die Beugungswelle in der Kirchhoffschen Theorie der Beugungserscheinungen, *Annalen der Physik* (4th ser.) 53(12) (1917): 257–278.

[26] Rubinowicz, A.: Zur Kirchhoffschen Beugungstheorie, *Annalen der Physik* (4th ser.) 73(5–6) (1924): 339–364.

[27] Rubinowicz, A.: *Die Beugungswelle in der Kirchhoffschen Theorie der Beugung*, 2nd ed., Berlin: Springer-Verlag, 1966.

[28] Sommerfeld, Arnold: Mathematische Theorie der Diffraction, *Mathematische Annalen* 47 (1896): 317–374.

[29] Sommerfeld, A. & Iris Runge, Anwendung der Vektorrechnung auf die Grundlagen der geometrischen Optik, *Annalen der Physik* (4th ser.) 35(7) (1911): 277–298.

[30] Sommerfeld, A., Die Greensche Funktion der Schwingungsgleichung, *Jahresberichte der Deutschen Mathematiker-Vereinigung* 21 (1912): 309–353.

[31] Sommerfeld, Arnold: *Optik. Vorlesungen über Theoretische Physik*, vol. 4, Frankfurt: Harri Deutsch, 1989.

[32] Sumbatyan, M. A. & A. Scalia: *Equations of Mathematical Diffraction Theory,* Boca Raton: Chapman & Hall/CRC, 2005.

[33] Synowiec, J.: Distributions — the evolution of a mathematical theory, *Historia Mathematica,* 10(2) (1983): 149–183.

[34] Taylor, M.E.: *Pseudodifferential Operators*, Princeton: Princeton University Press, 1981.

[35] Ufimtsev, P. Ya.: *Fundamentals of the Physical Theory of Diffraction*, Hoboken, New Jersey: Wiley-Interscience, 2007.

[36] Ufimtsev, P. Ya.: *Fundamentals of the Physical Theory of Diffraction*, 2nd ed., Hoboken, New Jersey: Wiley & Sons, 2014.

[37] Van Kampen, N. G.: The method of stationary phase and the method of Fresnel zones, *Physica* XXIV (1958): 437–444.

[38] Watson, G. N.: *A Treatise on the Theory of Bessel Functions*, 2nd ed., Cambridge: Cambridge University Press, 1995.

[39] Wong, R.: *Asymptotic Approximations of Integrals*, Philadelphia: SIAM, 2001.

Lecturer (Privatdozent) Dr. Ning Yan Zhu

Institute of Radio Frequency Technology

University of Stuttgart

Pfaffenwaldring 47

D-70550 Stuttgart, Germany

Email: zhu@ihf.uni-stuttgart.de

Gustav Robert Kirchhoff: *On the Theory of Light Rays*, presented to the Royal Prussian Academy of Sciences in 1882

Zur Theorie der Lichtstrahlen. *Sitzungsberichte der Königlich Preussischen Akademie der Wissenschaften zu Berlin*, 1882, part 2, pp. 641–669 [also publ. in *Annalen der Physik* **255** (= new ser. **18**) (1883), pp. 663–695]. Translated by Ann M. Hentschel.

The conclusions generally drawn from observations by Huyghens and Fresnel to explain the way light rays are formed, reflected and refracted as well as to explain diffraction phenomena lack rigor in many respects. It still does not seem possible today to develop a fully satisfactory theory on these subjects from the hypotheses of the undulatory theory; greater precision may, however, be given to these conclusions. I permit myself to present to the Academy arguments to this purpose, the substance of which I have been offering in my university lectures since a number of years. A few published articles by Messrs. Fröhlich[1] and Voigt[2] have meanwhile pursued the same goal with regard to diffraction phenomena.

---

[1][Orig. note 1, p. 641: J. Fröhlich, Einführung des Princips der Erhaltung der Energie und der Theorie der Diffraction,] Wiedemann's *Annalen* [*der Physik*, new ser.,] vol. 3, [1878] pp. 376–388; [Experimentaluntersuchungen über die Intensität des gebeugten Lichtes, ibid., pp. 568–581.]; [Die Bedeutung des Princips der Erhaltung der Energie in der Diffractionstheorie, ibid.,] vol. 6, [1879] pp. 414–431; and [Experimentaluntersuchungen über die Intensität des gebeugten Lichtes. II. Ibid.,] vol. 15, [1882] pp. 579–613, esp.] p. 592.

[2][Orig. note 2, p. 641: W. Voigt, Zur Fresnel'schen Theorie der Diffractionserscheinungen,] Wiedemann's *Annalen* [*der Physik*, new ser.,], vol. 3, [1878] pp. 532–568.

§1

It should be assumed that light constitutes transversal oscillations of the ether and that, with reference to these oscillations within the medium in which the moving light is observed, the ether acts like a rigid, elastic, isotropic and homogeneous body upon whose parts no other forces are acting than those caused by the relative displacements. If u, v, w are the components for the coordinate axes of the displacements of an ether particle having, in the equilibrium position, the coordinates x, y, z at time t, then each of these quantities satisfies the partial differential equation

$$\frac{\partial^2 \phi}{\partial t^2} = a^2 \Delta \phi, \tag{1}$$

where Δ means the sum of the second derivatives of x, y, z and a the light's propagation velocity. However, it is not permissible to set arbitrary solutions to this equation equal to u, v, w, because

$$\frac{\partial u}{\partial x} + \frac{\partial v}{\partial y} + \frac{\partial w}{\partial z} = 0$$

must also hold. If U, V, W are arbitrary solutions to this equation, then

$$u = \frac{\partial V}{\partial z} - \frac{\partial W}{\partial y}$$

$$v = \frac{\partial W}{\partial x} - \frac{\partial U}{\partial z} \tag{2}$$

$$w = \frac{\partial U}{\partial y} - \frac{\partial V}{\partial x}$$

correspond to one possible motion of the light; and conversely, for each motion of the light there are functions U, V, W satisfying these equations.[3] In the following let ϕ be understood to be one of the quantities U, V, W or u, v, w. If T is the oscillation period of the light, assumed to be homogeneous, then each of these 6 quantities is a linear, homogeneous

---

[3] [Orig. note 1, p. 642: A.] Clebsch, [Über die Reflexion an einer Kugelfläche,] in Borchardt's *Journal* [*für die reine und angewandte Mathematik*], vol. 61, [1862, pp. 195–262].

function of

$$\cos \frac{t}{T} 2\pi \quad \text{and} \quad \sin \frac{t}{T} 2\pi.$$

The arithmetic mean of the values should be taken as a measure of the light's intensity at point (x, y, z), which obtains

$$u^2 + v^2 + w^2$$

during time T; i.e., if one sets

$$u = \mathfrak{u} \cos \frac{t}{T} 2\pi + \mathfrak{u}' \sin \frac{t}{T} 2\pi$$

$$v = \mathfrak{v} \cos \frac{t}{T} 2\pi + \mathfrak{v}' \sin \frac{t}{T} 2\pi$$

$$w = \mathfrak{w} \cos \frac{t}{T} 2\pi + \mathfrak{w}' \sin \frac{t}{T} 2\pi,$$

then

$$\frac{1}{2} (\mathfrak{u}^2 + \mathfrak{u}'^2 + \mathfrak{v}^2 + \mathfrak{v}'^2 + \mathfrak{w}^2 + \mathfrak{w}'^2).$$

If the entire infinite space is filled with the considered medium; if a luminous point is situated in it at the position of point ι, whose coordinates are $x_\iota, y_\iota, z_\iota$; and if one denotes as r_ι the distance between the points (x, y, z) and $(x_\iota, y_\iota, z_\iota)$; the wavelength of the light as λ, i.e., the product aT; then the simplest assumption one can make about ϕ that is also permissible if one comprehends ϕ as one of the three quantities U, V, W is:

$$\phi = \frac{1}{r_\iota} \cos \left(\frac{r_\iota}{\lambda} - \frac{t}{T} \right) 2\pi. \tag{3}$$

From this expression of ϕ it is possible to derive a more general one that refers to the same case, by adding to it a constant factor, an additive constant to t, taking the derivative once or repeatedly for x_ι, y_ι, or z_ι, and summing over the thus-formed expressions. The result of this operation is substantially simplified if one introduces the assumption of fundamental importance in optics that wavelength λ may be regarded as infinitesimal. One thereby obtains, by taking into account only the terms of highest order,

$$\phi = \frac{D}{r_\iota} \cos \left(\frac{r_\iota}{\lambda} - \frac{t}{T} \right) 2\pi + \frac{D'}{r_\iota} \sin \left(\frac{r_\iota}{\lambda} - \frac{t}{T} \right) 2\pi, \tag{4}$$

where D and D' depend on $\frac{\partial r_\iota}{\partial x_\iota}, \frac{\partial r_\iota}{\partial y_\iota}, \frac{\partial r_\iota}{\partial z_\iota}$, or on $\frac{\partial r_\iota}{\partial x}, \frac{\partial r_\iota}{\partial y}, \frac{\partial r_\iota}{\partial z}$, which is the same thing, i.e., from the **direction** of line r_ι, but are otherwise constant. Then, according to (2), expressions of the same form also apply for u, v, w; if one designates as A, A', B, B', or C, C' the values for D and D' for the case where ϕ is set $= u, = v$ or $= w$, therefore letting these 6 variables signify quantities dependent on the direction of the line r_ι but are otherwise constant, then the intensity of the light at point (x, y, z) is

$$= \frac{1}{2r_\iota^2}(A^2 + A'^2 + B^2 + B'^2 + C^2 + C'^2).$$

Thus is expressed that this intensity is inversely proportional to the square of the distance from the luminous point, but varies in the direction of line r_ι in a manner defined by the motion at the luminous point.

A luminous point as the one considered should in the following considerations be assumed to be a light source so as to examine how the light emitted from it is modified by a foreign body brought into its proximity. An essential tool in this analysis will be a theorem that results in allowing Green's theorem to be applied to functions that satisfy the differential equation set for ϕ and that specifies and generalizes the so-called Huyghens theorem. Mr. Helmholtz has already derived it in his "Theory of Air Oscillations in Tubes with Open Ends"[4] and has shown its importance; the following section aims to develop this theorem in another way and in another form.

§2

Let $\mathfrak{U}$ and $\mathfrak{V}$ be two functions of x, y, z within a completely bounded space (that can also be composed of many separate parts), whose first derivatives of x, y, z are defined and continuous; $d\tau$ be one element of this space; ds one element of its surface (which can likewise be composed of separate parts); and N be the normal of ds directed toward the interior of the space; then according to Green's theorem

$$\int ds \left(\mathfrak{U}\frac{\partial \mathfrak{V}}{\partial N} - \mathfrak{V}\frac{\partial \mathfrak{U}}{\partial N} \right) = \int d\tau (\mathfrak{V}\Delta\mathfrak{U} - \mathfrak{U}\Delta\mathfrak{V}).$$

---

[4][Orig. note 1, p. 643: H. von Helmholtz, Theorie der Luftschwingungen in Röhren mit offenen Enden], Borchardt's *Journal [für die reine und angewandte Mathematik]*, vol. 57, [1860, pp. 11–72].

Here set $\mathfrak{U} = \phi$ and regarding $\mathfrak{V}$ initially assume that it also satisfies equation (1). One then obtains

$$\int ds \left(\phi \frac{\partial \mathfrak{V}}{\partial N} - \mathfrak{V} \frac{\partial \phi}{\partial N} \right) = \frac{1}{a^2} \int d\tau \left(\mathfrak{V} \frac{\partial^2 \phi}{\partial t^2} - \phi \frac{\partial^2 \mathfrak{V}}{\partial t^2} \right)$$

$$\text{or} \quad = \frac{1}{a^2} \frac{\partial}{\partial t} \int d\tau \left(\mathfrak{V} \frac{\partial \phi}{\partial t} - \phi \frac{\partial \mathfrak{V}}{\partial t} \right).$$

This equation is multiplied by dt and integrated over two values for time, one of which is negative, the other positive, and may be called $-t'$ and t''. A customary denotation thus yields

$$\int_{-t'}^{t''} dt \int ds \left(\phi \frac{\partial \mathfrak{V}}{\partial N} - \mathfrak{V} \frac{\partial \phi}{\partial N} \right) = \frac{1}{a^2} \left[\int d\tau \left(\mathfrak{V} \frac{\partial \phi}{\partial t} - \phi \frac{\partial \mathfrak{V}}{\partial t} \right) \right]_{-t'}^{t''}. \quad (5)$$

Now, let

$$\mathfrak{V} = \frac{F(r_o + at)}{r_o},$$

where r_o signifies the distance of the point (x, y, z) from an arbitrarily chosen point o and F is a function that vanishes for every finite, positive or negative value of its argument, is never negative and satisfies the condition that

$$\int F(\zeta) d\zeta = 1, \quad (6)$$

if the integration is extended over one finite negative value of ζ up to a finite positive one.

Now, given a completely bounded space that is filled with homogeneous ether and free of luminous points; let s be its surface and ds one element of it. The point o is assumed to be in the interior of this space and equation (5) is applied to the space remaining **after** excluding an infinitesimal sphere, the center of which is the point o. Let dS be one element of the surface of this sphere. Choose the size of t' so that

$$r_o - at'$$

the largest value that r_o gets on surface s — hence throughout the imagined space — is negative and finite; under this condition, only values for $\mathfrak{V}$ and $\frac{\partial \mathfrak{V}}{\partial t}$ occur on the right-hand side of equation (5) for which $r_o + at$ is **finite**, positive or negative and which consequently vanish. Equation (5) thus yields

$$\int_{-t'}^{t''} dt \int ds \left(\phi \frac{\partial \mathfrak{V}}{\partial N} - \mathfrak{V} \frac{\partial \phi}{\partial N} \right) + \int_{-t'}^{t''} dt \int dS \left(\phi \frac{\partial \mathfrak{V}}{\partial N} - \mathfrak{V} \frac{\partial \phi}{\partial N} \right) = 0. \quad (7)$$

The second of these integrals can be evaluated. If one denotes as R the radius of the infinitesimal sphere to which it refers and in the calculation of the term multiplied by dS omits the infinitesimal product of R^2, one may set

$$\frac{\partial \mathfrak{V}}{\partial N} = -\frac{1}{R^2} F(at), \quad \mathfrak{V} = 0,$$

hence,

$$\int dS \left(\phi \frac{\partial \mathfrak{V}}{\partial N} - \mathfrak{V} \frac{\partial \phi}{\partial N} \right) = -4\pi \phi_o F(at),$$

where ϕ_o signifies the value of ϕ for the point o. As, furthermore, $F(at)$ differs from zero by only infinitesimal values of t and, according to equation (6),

$$\int_{-t'}^{t''} dt F(at) = \frac{1}{a},$$

the second term of equation (7) becomes

$$-\frac{4\pi}{a} \phi_o(o),$$

where $\phi_o(o)$ denotes the value of ϕ_o for $t = 0$. The first term also allows the integration to be taken for t with the aid of equation (6). One initially has

$$a \int_{-t'}^{t''} dt \mathfrak{V} \frac{\partial \phi}{\partial N} = a \int_{-t'}^{t''} dt \frac{F(r_o + at)}{r_o} \frac{\partial \phi}{\partial N} = \frac{1}{r_o} \frac{\partial \phi}{\partial N},$$

where, after taking the derivative,

$$t = -\frac{r_o}{a}$$

must be set in $\frac{\partial \phi}{\partial N}$. Putting

$$\frac{\partial \phi}{\partial N} = f(t), \quad (8)$$

the expression thus becomes

$$\frac{1}{r_o} f \left(-\frac{r_o}{a} \right).$$

Furthermore,

$$\frac{\partial \mathfrak{V}}{\partial N} = \frac{\partial \frac{F(r_o+at)}{r_o}}{\partial N} = \frac{\partial \frac{1}{r_o}}{\partial N} F(r_o + at) + \frac{1}{r_o} \frac{\partial r_o}{\partial N} \frac{1}{a} \frac{\partial F(r_o + at)}{\partial t}$$

holds and therefore

$$a \int_{-t'}^{t''} dt \phi \frac{\partial \mathfrak{V}}{\partial N} = \frac{\partial \frac{1}{r_o}}{\partial N} \phi \left(-\frac{r_o}{a} \right) + \frac{1}{r_o} \frac{\partial r_o}{\partial N} \int_{-t'}^{t''} \phi \frac{\partial F(r_o + at)}{\partial t} dt,$$

where $\phi\left(-\frac{r_o}{a}\right)$ means the value of ϕ for $t = -\frac{r_o}{a}$. Reformulating the last integral through partial integration and considering that function F vanishes for every finite value of its argument, one finds the same expression

$$= \frac{\partial \frac{1}{r_o}}{\partial N} \phi \left(-\frac{r_o}{a} \right) - \frac{1}{a} \frac{1}{r_o} \frac{\partial r_o}{\partial N} \frac{\partial \phi}{\partial t},$$

where, in $\frac{\partial \phi}{\partial t}$ one should likewise set $t = -\frac{r_o}{a}$. Substituting these results in equation (7) and at the same time shifting the time origin so that the previous time origin becomes t, one gets

$$4\pi \phi_o(t) = \int ds \left\{ \frac{\partial \frac{1}{r_o}}{\partial N} \phi \left(t - \frac{r_o}{a} \right) - \frac{1}{a} \frac{1}{r_o} \frac{\partial r_o}{\partial N} \frac{\partial \phi \left(t - \frac{r_o}{a} \right)}{\partial t} - \frac{1}{r_o} f \left(t - \frac{r_o}{a} \right) \right\}.$$
(9)

The first two terms of the expression multiplied here by ds can be reduced to one

$$\frac{\partial}{\partial N} \frac{\phi \left(t - \frac{r_o}{a} \right)}{r_o},$$

whereby the derivative should be taken such that only r_o is regarded as variable, while the quantities upon which $\phi(t)$ depends should retain the values belonging to them in element ds. One accordingly has

$$4\pi \phi_o(t) = \int ds \, \Omega,$$
(10)

where

$$\Omega = \frac{\partial}{\partial N} \frac{\phi(t - \frac{r_o}{a})}{r_o} - \frac{f(t - \frac{r_o}{a})}{r_o},$$
(11)

and where function f is defined by equation (8).

The conclusion from this is that the motion of the ether in the space enclosed in surface s can be regarded as caused by a layer of luminous points on surface s, because each one of the two terms of which Ω is composed may be described as corresponding to a luminous point situated at the location of ds.

The following consideration proves that under a particular condition that will henceforth always be presumed satisfied, equation (10) also holds when the luminous points lie within the space enclosed by surface s and point o is situated outside of it; the normal N must then just be pointed outwards. In this case one applies equation (10) to the space limited inwards by surface s and outwards by an infinitely large spherical surface whose element may be denoted as dS. One thus obtains

$$4\pi\phi_o(t) = \int ds\,\Omega + \int dS\,\Omega.$$

Now assume that up to a certain finite value for time, the state of rest prevails everywhere so for infinitely large negative values of t, hence also the infinitely large sphere, $\phi(t)$ and $f(t)$ vanish. If one chooses point o in finitude and takes into regard only finite values of time, then Ω vanishes for every element dS, because here $t - \frac{r_o}{a}$ is negatively infinite; one thus obtains equation (10). The constraint that point o should lie within finitude and time should be finite is consequently merely an apparent one: whatever may be the position of point o and the value of t, the sphere's radius can be chosen large enough to maintain the consideration's validity.

If equation (10) is applied to two closed surfaces that share one part in common and both enclose point o, but not the luminous point — or, indeed, the luminous points, but not point o — and the results thus obtained are subtracted from each other, one sees that the integral $\int ds\,\Omega$, extended over a closed surface that encloses neither the luminous points nor point o, vanishes. It also vanishes for a closed surface that encloses point o and the luminous points, as one realizes when equation (10) is formed for two closed surfaces that have a common part and one surface of which encloses point o and not the luminous points; the other enclosing the luminous points and not point o.

The application of equation (10) in the foregoing to the problem described at the end of the previous section is obvious. Imagine in the homogeneous ether filling the infinite space a luminous point ι; relate function ϕ^* to the motion it exhibits. If a foreign body is brought into the space, the motion is altered; ϕ^* is thus made into ϕ; the task involved is to determine ϕ for a given point o that lies outside of the body. Let ds be one element of the body's surface, dS one element of an infinitesimal spherical surface circumscribed around the luminous point; then according to equation (10),

$$4\pi\phi_o = \int dS\,\Omega + \int ds\,\Omega.$$

The first of these two integrals has an easily indicatable value. The change in the motion of element dS caused by the introduction of the body is not infinitely large (excluding one particular special case) and, since the spherical surface belonging to dS is infinitesimal, its influence on the integral's value is infinitesimal. So in this integral ϕ^* can be substituted for ϕ, whereby according to equation (10), this becomes $= 4\pi\phi_o^*$, if ϕ_o^* denotes the value of ϕ^* at point o. One therefore has

$$4\pi\phi_o = 4\pi\phi_o^* + \int ds\,\Omega. \tag{12}$$

According to this equation ϕ_o can be calculated generally if ϕ^* and the values of ϕ and $\frac{\partial\phi}{\partial N}$ for the body's surface are known.

§3

For the later considerations it is necessary to know the value that the integral $\int ds\,\Omega$ has, extended over a **bounded** surface under certain conditions. This value shall now be derived. For this it should be presumed that the wavelength is infinitesimal, that ϕ originates from a luminous point ι, hence has the expression indicated in equation (4), that for no finite part of surface s over which the integral extends, nor for its boundary, is $r_\iota + r_o$ constant or constant up to an infinitesimally small quantity, and finally, that the straight line connecting points ι and o does not go through the surface's boundary or pass infinitely close by it. It will be proved that then the mentioned integral will vanish in the case where the straight line connecting ι to o does not intersect surface s. The calculation will yield that when such an intersection takes place, the integral is $= \pm 4\pi\phi_o$, where the upper or the lower sign holds, depending on whether the angle formed by the normal N at the intersecting point with the line drawn from ι to o is acute or obtuse; which, provided the first assumption is proven, already follows from equation (10).

First assume the expression given in (3) for ϕ, thus set

$$\phi = \frac{1}{r_\iota} \cos\left(\frac{r_\iota}{\lambda} - \frac{t}{T}\right) 2\pi;$$

then results

$$\frac{\partial}{\partial N} \frac{1}{r_o} \phi\left(t - \frac{r_o}{a}\right) = -\frac{1}{r_\iota r_o^2} \frac{\partial r_o}{\partial N} \cos\left(\frac{r_\iota + r_o}{\lambda} - \frac{t}{T}\right) 2\pi$$

$$- \frac{2\pi}{r_\iota r_o \lambda} \frac{\partial r_o}{\partial N} \sin\left(\frac{r_\iota + r_o}{\lambda} - \frac{t}{T}\right) 2\pi,$$

furthermore, according to equation (8)

$$\frac{1}{r_o} f\left(t - \frac{r_o}{a}\right) = -\frac{1}{r_\iota^2 r_o} \frac{\partial r_\iota}{\partial N} \cos\left(\frac{r_\iota + r_o}{\lambda} - \frac{t}{T}\right) 2\pi$$

$$- \frac{2\pi}{r_\iota r_o \lambda} \frac{\partial r_\iota}{\partial N} \sin\left(\frac{r_\iota + r_o}{\lambda} - \frac{t}{T}\right) 2\pi$$

and therefore, according to equation (11),

$$\Omega = \frac{1}{r_\iota r_o} \left(\frac{1}{r_\iota} \frac{\partial r_\iota}{\partial N} - \frac{1}{r_o} \frac{\partial r_o}{\partial N}\right) \cos\left(\frac{r_\iota + r_o}{\lambda} - \frac{t}{T}\right) 2\pi$$

$$+ \frac{2\pi}{r_\iota r_o \lambda} \left(\frac{\partial r_\iota}{\partial N} - \frac{\partial r_o}{\partial N}\right) \sin\left(\frac{r_\iota + r_o}{\lambda} - \frac{t}{T}\right) 2\pi. \tag{13}$$

In order to find the mentioned integral with this value for Ω, one proceeds from the following theorem.

If $F(\zeta)$ describes a function of ζ that is continuous in the interval in which ζ increases from ζ_o to ζ', and δ is a constant, then the integral

$$\int_{\zeta_o}^{\zeta'} \frac{dF}{d\zeta} \sin(k\zeta + \delta) d\zeta, \tag{14}$$

vanishes if k becomes infinitely large.

The correctness of this theorem follows from considerations entirely similar to the ones Dirichlet has made in his analyses on the Fourier series with reference to a similar integral. One resolves the integral in parts such that within each $\frac{dF}{d\zeta}$ there is neither a change in sign nor in the tendency, from diminishing to increasing or vice versa; each of these parts (assumed to be finite in number) are proved to vanish if k increases to infinity by continuing to decompose the parts so that all values of ζ for which $\sin(k\zeta+\delta) = 0$ holds occur as intermediary boundaries and by using the differences denotable as absolute values of these parts.

From this theorem the following easily results.

When the function $F(\zeta)$ has the property that its first derivative is continuous in the interval from $\zeta = \zeta_o$ to $\zeta = \zeta'$, then results for $k = \infty$

$$k \int_{\zeta_o}^{\zeta'} \frac{dF}{d\zeta} \sin(k\zeta + \delta)d\zeta = - \left[\frac{dF}{d\zeta} \cos(k\zeta + \delta)\right]_{\zeta_o}^{\zeta'}. \tag{15}$$

In fact, the left half of this equation becomes, through partial integration,

$$= - \left[\frac{dF}{d\zeta} \cos(k\zeta + \delta)\right]_{\zeta_o}^{\zeta'} + \int_{\zeta_o}^{\zeta'} \frac{d^2 F}{d\zeta^2} \cos(k\zeta + \delta)d\zeta;$$

the new integral appearing here is of the form of integral (14), however, so it vanishes if k increases into infinity.

Now imagine a continuously curved, completely bounded surface s, whose element is ds, denote r_ι and r_o as the distances of this element from two fixed points ι and o, set

$$\zeta = r_\iota + r_o,$$

describe as G a continuously varying function of the position of ds, as δ a constant, and examine the value that the integral

$$\int G \sin(k\zeta + \delta)ds \tag{16}$$

takes when k gets infinitely large.

For this purpose imagine the surfaces whose equations are

$$\zeta = \text{const.},$$

thus the rotational ellipsoids whose foci are the points ι and o, and their intersecting lines with surface s; then set

$$F(\zeta) = \pm \int Gds, \tag{17}$$

where the integration is to be extended over the part of surface s that lies between the two intersection lines of which one corresponds to the variable value ζ and the other is an arbitrarily chosen fixed value Z and where the $+$ sign is valid if $\zeta > Z$, the $-$ sign if $\zeta < Z$. With this definition, when a positive $d\zeta$ is chosen,

$$\frac{dF}{d\zeta}d\zeta = \int Gds, \tag{18}$$

where the integration should be extended over the part of surface s that lies between the two intersection lines corresponding to the values ζ and $\zeta + d\zeta$. If ζ_o is the smallest value for ζ and ζ' is the largest on surface s, then accordingly integral (16) is

$$= \int_{\zeta_o}^{\zeta'} \frac{dF}{d\zeta} \sin(k\zeta + \delta)d\zeta,$$

hence $=$ integral (14); it therefore vanishes for $k = \infty$, if $F(\zeta)$ is continuous on surface s, i.e., if no constant value for ζ occurs on any finite part of surface s.

We shall now view the expression with the same definitions for the symbols

$$k \int G\sin(k\zeta + \delta)ds. \tag{19}$$

This is

$$= k \int_{\zeta_o}^{\zeta'} \frac{dF}{d\zeta} \sin(k\zeta + \delta)d\zeta,$$

hence the same as the left-hand side of equation (15). It is consequently also the same as the right-hand side of the same equation for $k = \infty$, provided $\frac{dF}{d\zeta}$ within surface s as defined by equation (18) is continuous. This derivative is discontinuous as soon as ζ is constant for a finite part of the boundary of s; if this case is excluded, a discontinuity can only occur if $d\zeta$ vanishes for a point on the surface. We shall specially analyze what then takes place. Otherwise equation (15) is valid; and from it follows furthermore that expression (19) vanishes. Under the preconditions made, not only the largest value for ζ but also the smallest value occurs in one or some points of the **boundary** of s; and for each one of such points, the integral $\int Gds$ that has to be calculated in order to find out the corresponding $\frac{dF}{d\zeta}$ pursuant to (18), is infinitesimal of higher order than $d\zeta$; therefore this $\frac{dF}{d\zeta}$ vanishes.

Now the value for expression (19) must be sought for the case that $d\zeta$ vanishes for a point on surface s. Let this be for the point (x, y, z) and

$g(x, y, z) = 0$ be the equation of this surface; then holds

$$\frac{\partial r_\iota}{\partial x} + \frac{\partial r_o}{\partial x} = L\frac{\partial g}{\partial x}$$

$$\frac{\partial r_\iota}{\partial y} + \frac{\partial r_o}{\partial y} = L\frac{\partial g}{\partial y}$$

$$\frac{\partial r_\iota}{\partial z} + \frac{\partial r_o}{\partial z} = L\frac{\partial g}{\partial z},$$

where L signifies an undefined factor. If $\alpha_\iota, \beta_\iota, \gamma_\iota$, $\alpha_o, \beta_o, \gamma_o$ and α, β, γ denote the cosines of the angles that the coordinate axes form with the line that is drawn from point ι to point (x, y, z); with the line that is drawn from point o to point (x, y, z); and with a normal N of surface s at this point, then these equations may be written:

$$\alpha_\iota + \alpha_o = M\alpha$$
$$\beta_\iota + \beta_o = M\beta \tag{20}$$
$$\gamma_\iota + \gamma_o = M\gamma,$$

where M denotes a new factor. From these results, firstly, that the lines r_ι, r_o and N lie on **one** plane; then also follows

$$M(\alpha\alpha_\iota + \beta\beta_\iota + \gamma\gamma_\iota) = M(\alpha\alpha_o + \beta\beta_o + \gamma\gamma_o),$$

and this equation states that either $M = 0$, i.e., $\alpha_o = -\alpha_\iota$, $\beta_o = -\beta_\iota$, $\gamma_o = -\gamma_\iota$, therefore that point (x, y, z) lies between the points ι and o on their straight connecting line, or the directions $(\alpha_\iota, \beta_\iota, \gamma_\iota)$ and $(\alpha_o, \beta_o, \gamma_o)$ form the same angles with the direction N. In the second case the lines r_ι and r_o must lie on opposite sides of the normal N, if they do not coincide with it or its extension; for, $\alpha_o = \alpha_\iota$, $\beta_o = \beta_\iota$, $\gamma_o = \gamma_\iota$ do not satisfy equations (20) unless r_ι and r_o coincide with N or the extension of N.

The meanings of the symbols x, y, z are now altered and (x, y, z) denotes a variable point of surface s with reference to a coordinate system whose origin is the prior point (x, y, z) and whose z-axis is the normal N. Furthermore, the dimensions of surface s should be assumed to be infinitesimal (but infinitely large against $\frac{1}{k}$); it suffices under this assumption to calculate integral (19), because, as has been proven, its value is not changed by the addition of new parts to surface s. The equation of surface s then is

$$z = a_{11}x^2 + 2a_{12}xy + a_{22}y^2, \tag{21}$$

where a_{11}, a_{12}, a_{22} are constants, and at the same time

$$ds = dx\, dy.$$

In order to find the intersecting lines of surface s with the surfaces $\zeta = $ const., the expression for ζ must now be formulated and expanded by the powers of x and y. Let x_o, y_o, z_o be the coordinates of the point o and

$$\rho_o = \sqrt{x_o^2 + y_o^2 + z_o^2};$$

then

$$r_o = \sqrt{(x - x_o)^2 + (y - y_o)^2 + (z - z_o)^2}$$

or

$$r_o = \sqrt{\rho_o^2 - 2xx_o - 2yy_o - 2zz_o + x^2 + y^2 + z^2}.$$

If x and y are described as infinitesimal of first order and r_o is expanded using equation (21) up to and including quantities of second order, then results

$$r_o = \rho_o - \frac{xx_o + yy_o}{\rho_o} - \frac{a_{11}x^2 + 2a_{12}xy + a_{22}y^2}{\rho_o}z_o$$

$$+ \frac{x^2 + y^2}{2\rho_o} - \frac{(xx_o + yy_o)^2}{2\rho_o^3},$$

or, as the quantities $\alpha_o, \beta_o, \gamma_o$ occurring in equation (20) satisfy the equations

$$\frac{x_o}{\rho_o} = -\alpha_o, \quad \frac{y_o}{\rho_o} = -\beta_o, \quad \frac{z_o}{\rho_o} = -\gamma_o,$$

thus

$$r_o = \rho_o + \alpha_o x + \beta_o y + (a_{11}x^2 + 2a_{12}xy + a_{22}y^2)\gamma_o$$

$$+ \frac{1}{2\rho_o}(x^2(1 - \alpha_o^2) - 2xy\alpha_o\beta_o + y^2(1 - \beta_o^2)).$$

Appropriately setting

$$\rho_\iota = \sqrt{x_\iota^2 + y_\iota^2 + z_\iota^2},$$

one likewise finds

$$r_\iota = \rho_\iota + \alpha_\iota x + \beta_\iota y + (a_{11}x^2 + 2a_{12}xy + a_{22}y^2)\gamma_\iota$$

$$+ \frac{1}{2\rho_\iota}(x^2(1 - \alpha_\iota^2) - 2xy\alpha_\iota\beta_\iota + y^2(1 - \beta_\iota^2)).$$

For the chosen coordinate system, however, $\alpha = 0$ and $\beta = 0$, and therefore according to (20)

$$\alpha_\iota + \alpha_o = 0, \quad \beta_\iota + \beta_o = 0.$$

One therefore has

$$\zeta = A_o + A_{11}x^2 + 2A_{12}xy + A_{22}y^2,$$

where

$$A_o = \rho_\iota + \rho_o$$

$$A_{11} = a_{11}(\gamma_\iota + \gamma_o) + \frac{1 - \alpha_\iota^2}{2\rho_\iota} + \frac{1 - \alpha_o^2}{2\rho_o}$$

$$A_{12} = a_{12}(\gamma_\iota + \gamma_o) - \frac{\alpha_\iota \beta_\iota}{2\rho_\iota} - \frac{\alpha_o \beta_o}{2\rho_o}$$

$$A_{22} = a_{22}(\gamma_\iota + \gamma_o) + \frac{1 - \beta_\iota^2}{2\rho_\iota} + \frac{1 - \beta_o^2}{2\rho_o}.$$

$$(22)$$

The intersection curves of the surfaces $\zeta = $ const. with surface s are accordingly similar as well as similarly lying conic sections whose common center point is the origin of the coordinates. Their equation, with reference to the main axes, would be

$$\zeta - A_o = \mu_1 x^2 + \mu_2 y^2,$$

i.e., let μ_1 and μ_2 be the (always real) roots of the quadratic equation

$$(A_{11} - \mu)(A_{22} - \mu) - A_{12}^2 = 0. \qquad (23)$$

If μ_1 and μ_2 have the same sign, then the conic sections are ellipses: A_o is the minimum of ζ if μ_1 and μ_2 are positive, the maximum if both these quantities have a negative sign. In the first case, the surface of the ellipse, which corresponds to a value of ζ,

$$= \frac{\pi(\zeta - A_o)}{\sqrt{\mu_1 \mu_2}},$$

in the second

$$= \frac{\pi(A_o - \zeta)}{\sqrt{\mu_1 \mu_2}},$$

where the root taken should be positive, as generally here the root of a positive quantity should be interpreted as positive. According to equation (17), therefore, if the quantity designated there as Z is chosen to be $= A_o$ for values of ζ whose corresponding ellipses lie entirely within surface s, in both cases

$$F(\zeta) = G\frac{\pi(\zeta - A_o)}{\sqrt{\mu_1\mu_2}},$$

holds, where G refers to the point $(x = 0, y = 0)$, thus

$$\frac{dF}{d\zeta} = G\frac{\pi}{\sqrt{\mu_1\mu_2}}.$$

If no part of the boundary of s coincides with one of the ellipses, then $\frac{dF}{d\zeta}$ is continuous on this surface and $= 0$ for the second boundary value that ζ reaches here. From this, expression (19) for $k = \infty$ is, if μ_1 and μ_2 are positive,

$$= G\frac{\pi}{\sqrt{\mu_1\mu_2}}\cos(kA_o + \delta), \tag{24}$$

and, if μ_1 and μ_2 are negative,

$$= -G\frac{\pi}{\sqrt{\mu_1\mu_2}}\cos(kA_o + \delta). \tag{25}$$

The calculation becomes less simple if μ_1 and μ_2 have opposite signs, so the conic sections are hyperbolas; in which case $\frac{dF}{d\zeta}$ is discontinuous for $\zeta = A_o$. One selects here the main axes as the coordinate axes and gives surface s a defined shape, namely that of a rectangle whose sides are parallel to the main axes and have the equations

$$x = \pm a, \quad y = \pm b.$$

The corners should lie on the asymptotes, so it should be

$$a\sqrt{\mu_1} = b\sqrt{-\mu_2} = c,$$

where μ_1 is positive, μ_2 is negative, and c is positive. The real main axis of a hyperbola corresponding to the value of ζ then coincides with the x-axis if $\zeta - A_o$ is positive, and with the y-axis if $\zeta - A_o$ is negative. If one again sets the quantity $Z = A_o$, defined by equation (17), one therefore obtains for $\zeta > A_o$

$$F(\zeta) = G \left\{ 2ab - \frac{4}{\sqrt{-\mu_2}} \int_{\sqrt{\frac{\zeta - A_o}{\mu_1}}}^{a} \sqrt{\mu_1 x^2 - \zeta + A_o} \, dx \right\},$$

where G again refers to the point $(x = 0, y = 0)$. From this follows

$$\frac{dF}{d\zeta} = G \frac{2}{\sqrt{-\mu_2}} \int_{\sqrt{\frac{\zeta - A_o}{\mu_1}}}^{a} \frac{dx}{\sqrt{\mu_1 x^2 - \zeta + A_o}},$$

or, as

$$\int_1^a \frac{dz}{\sqrt{z^2 - 1}} = \lg(z + \sqrt{z^2 - 1}),$$

$$\frac{dF}{d\zeta} = G \frac{2}{\sqrt{-\mu_1 \mu_2}} \lg \frac{c + \sqrt{c^2 - \zeta + A_o}}{\sqrt{\zeta - A_o}}.$$

One likewise obtains for $\zeta < A_o$

$$\frac{dF}{d\zeta} = G \frac{2}{\sqrt{-\mu_1 \mu_2}} \lg \frac{c + \sqrt{c^2 + \zeta - A_o}}{\sqrt{A_o - \zeta}}.$$

Considering that the smallest value for ζ takes place at the points $(x = 0, y = \pm b)$ and is $= A_o - c^2$, whereas the largest one occurs at the points $(x = \pm a, y = 0)$ and is $= A_o + c^2$, we get expression (19)

$$= G \frac{2}{\sqrt{-\mu_1 \mu_2}} k \left\{ \int_{A_o - c^2}^{A_o} \lg \frac{c + \sqrt{c^2 + \zeta - A_o}}{\sqrt{A_o - \zeta}} \sin(k\zeta + \delta) d\zeta \right.$$

$$\left. + \int_{A_o}^{A_o + c^2} \lg \frac{c + \sqrt{c^2 - \zeta + A_o}}{\sqrt{\zeta - A_o}} \sin(k\zeta + \delta) d\zeta \right\}.$$

If one puts

$$A_o - \zeta = \xi,$$

in the first of these two integrals,

$$\zeta - A_o = \xi$$

in the second, then this expression becomes

$$= G\frac{2}{\sqrt{-\mu_1\mu_2}}k\int_0^{c^2} \lg\frac{c+\sqrt{c^2-\xi}}{\sqrt{\xi}}(\sin(k\xi+kA_o+\delta)-\sin(k\xi-kA_o-\delta))d\xi,$$

or

$$= G\frac{4}{\sqrt{-\mu_1\mu_2}}k\sin(kA_o+\delta)\int_0^{c^2}\lg\frac{c+\sqrt{c^2-\xi}}{\sqrt{\xi}}\cos k\xi d\xi.$$

But now,

$$k\int_0^{c^2}\lg\frac{c+\sqrt{c^2-\xi}}{\sqrt{\xi}}\cos k\xi d\xi$$

$$=\left[\sin k\xi\lg\frac{c+\sqrt{c^2-\xi}}{\sqrt{\xi}}\right]_{\xi=0}^{\xi=c^2}-\int_0^{c^2}\sin k\xi\frac{d}{d\xi}\lg(c+\sqrt{c^2-\xi})d\xi$$

$$+\frac{1}{2}\int_0^{c^2}\frac{\sin k\xi}{\xi}d\xi.$$

The first of these 3 terms is equal to zero for any value of k because the expression within the parentheses vanishes for $\xi=c^2$ as well as for $\xi=0$; the second is of the form of expression (14) and thus vanishes for $k=\infty$, as $\lg(c+\sqrt{c^2-\xi})$ is continuous also for $\xi=c^2$, although its derivative becomes infinite; the third, finally, is, for $k=\infty$,

$$=\frac{1}{2}\int_0^\infty\frac{\sin u\,du}{u}=\frac{\pi}{4}.$$

The sought value for expression (19) is therefore, if μ_1 and μ_2 are of opposite sign,

$$= G\frac{\pi}{\sqrt{-\mu_1\mu_2}}\sin(kA_o+\delta). \tag{26}$$

In the further discussion of expressions (24), (25) and (26), make use of the circumstance that as μ_1 and μ_2 are the roots of equation (23),

$$\mu_1\mu_2 = A_{11}A_{22}-A_{12}^2, \tag{27}$$

where A_{11}, A_{12}, A_{22} have the values indicated in (22).

As has been concluded from equations (20), the presently performed considerations relate to two cases: the first of these is that surface s is intersected by the straight line connecting points ι and o; the second is that there is a point on surface s that has the property that the lines drawn from it to points ι and o form equal angles with the normal of surface s and lie on **one** plane with it. The first of these cases will be examined here further. In this case,

$$\alpha_\iota + \alpha_o = 0, \quad \beta_\iota + \beta_o = 0, \quad \gamma_\iota + \gamma_o = 0;$$

equations (22) thus yield

$$A_{11} = \frac{1}{2}\left(\frac{1}{\rho_\iota} + \frac{1}{\rho_o}\right)(1 - \alpha_1^2)$$

$$A_{12} = -\frac{1}{2}\left(\frac{1}{\rho_\iota} + \frac{1}{\rho_o}\right)\alpha_\iota\beta_\iota$$

$$A_{22} = \frac{1}{2}\left(\frac{1}{\rho_\iota} + \frac{1}{\rho_o}\right)(1 - \beta_1^2);$$

and pursuant to (27)

$$\mu_1\mu_2 = \frac{1}{4}\left(\frac{1}{\rho_\iota} + \frac{1}{\rho_o}\right)\gamma_1^2.$$

The roots of equation (23), μ_1 and μ_2, are

$$\frac{1}{2}\left(\frac{1}{\rho_\iota} + \frac{1}{\rho_o}\right) \quad \text{and} \quad \frac{1}{2}\left(\frac{1}{\rho_\iota} + \frac{1}{\rho_o}\right)\gamma_\iota^2,$$

hence both are positive; that is why expression (19) is equivalent to expression (24); therefore it is

$$= \pm G2\pi \frac{\rho_\iota\rho_o}{\rho_\iota + \rho_o}\frac{1}{\gamma_\iota}\cos(k(\rho_\iota + \rho_o) + \delta), \tag{28}$$

where the positive or negative sign is chosen, depending on whether γ_ι is positive or negative.

For these considerations on expression (19), δ is assumed to be a constant; but they also apply if δ, like G, varies continuously with the location of ds; then, in the expressions (24), (25), (26) and (28), δ as well as G have to be related to the point $(x = 0, y = 0)$. This is understood if one considers that for a variable δ the integral (19) can be broken down by the formula

$$\sin(k\zeta + \delta) = \cos\delta \sin k\zeta + \sin\delta \cos k\zeta$$

into the sum of two integrals of the same form in which δ has the constant values 0 and $\frac{\pi}{2}$.

By means of the results reached it is now easy to prove the assertion stated at the beginning of this section with regard to the integral $\int ds\,\Omega$.

First let Ω have the value indicated in (13), hence ϕ in (3); set

$$\frac{2\pi}{\lambda} = k; \quad \text{and} \quad -\frac{t}{T}2\pi = \delta;$$

one then sees that the part of the mentioned integral stemming from the first term of Ω vanishes and that the part that yields the second term of Ω is equal to zero, if there is no point on surface s of the type that have straight connecting lines to points ι and o forming the same angle to the normal of the surface and lying on **one** plane with it and if the surface is not intersected by the line connecting points ι and o. For, if the first of these conditions is not fulfilled, the integral in question itself vanishes too; since, in order to find its value, one has to set in expression (24), (25) or (26) for G the value that

$$\frac{1}{r_\iota r_o}\left(\frac{\partial r_\iota}{\partial N} - \frac{\partial r_o}{\partial N}\right) \tag{29}$$

assumes at the indicated point and that value is equal to zero, since $\frac{\partial r_\iota}{\partial N}$ and $\frac{\partial r_o}{\partial N}$ are the cosines of the angles that are supposed to be the same as each other. That is why $\int ds\,\Omega$ only does not vanish when surface s is intersected by the line connecting points ι and o. Expression (28) in this case yields its value if one assigns G with the value that expression (29) has at the intersection point. If the orientation of N that occurs in (13) is allowed to coincide with the direction of the z-axis, which γ_ι refers to in (28), then

$$\frac{\partial r_\iota}{\partial N} = \gamma_\iota, \quad \frac{\partial r_o}{\partial N} = -\gamma_\iota$$

and therefore the value of (29)

$$= \frac{2\gamma_\iota}{\rho_\iota \rho_o},$$

thus

$$\int ds\,\Omega = \pm\frac{4\pi}{\rho_\iota + \rho_o}\cos\left(\frac{\rho_\iota + \rho_o}{\lambda} - \frac{t}{T}\right)2\pi$$

or

$$= \pm 4\pi\phi_o,$$

where the positive or negative signs hold, depending on whether γ_ι is positive or negative, i.e., depending on whether the normal N forms an acute or obtuse angle with the line drawn from ι to o.

Hereby the statement under discussion is proven for the case where ϕ has the value indicated by equation (3); it stays correct if one transgresses from this equation to the more general equation (4) in the manner indicated there.

§4

In order to be able to draw conclusions from equation (12), it is necessary to examine the values of ϕ and $\frac{\partial \phi}{\partial N}$ at the body's surface that the equation presupposes.

If plane light waves traveling through a transparent medium fall on a plane that is the boundary to a second medium, then reflected and refracted planar waves form. Their formation and the directions they have, according to experience, can be regarded as a consequence of the validity of linear, homogeneous equations with constant coefficients in-between the displacements of the ether particles at the boundary in both media and their derivatives. Let ϕ_e refer to the incident light, ϕ_r to the reflected light at point (ξ, η, ζ); for the first medium let $\zeta < 0$, for the second $\zeta > 0$ and

$$\phi_e = A \cos \left(\frac{l\xi + m\eta + n\zeta}{\lambda} - \frac{t + \alpha}{T} \right) 2\pi,$$

where l, m, n signify the cosines of the angles that the coordinate axes form with the wave normal of the direction in which the incident light is propagating. Then

$$\phi_r = cA \cos \left(\frac{l\xi + m\eta - n\zeta}{\lambda} - \frac{t + \alpha + \gamma}{T} \right) 2\pi,$$

holds, where c and γ are constants whose values depend on the definition of the symbol ϕ, the angle of incidence, the state of polarization of the incident light and the nature of both media. For $\zeta = 0$ one therefore has, if one applies the symbols $\phi_e(t)$ and $\phi_r(t)$ as equivalent to ϕ_e and ϕ_r,

$$\phi_r(t) = c\phi_e(t + \gamma)$$

and

$$\frac{\partial \phi_r(t)}{\partial \zeta} = -c \frac{\partial \phi_e(t + \gamma)}{\partial \zeta}, \tag{30}$$

of which equations the second can also be written as

$$\frac{\partial \phi_r(t)}{\partial N} = -c \frac{\partial \phi_e(t + \gamma)}{\partial N}, \tag{30}$$

if N, as earlier, signifies the boundary's normal turned toward the interior of the first medium.

If waves of different orientations are present at the same time in the incident light such that both ϕ_e and ϕ_r are sums of expressions that happen to be equivalent to these symbols, then corresponding equations exist for the component parts of these sums.

These laws can be applied to the case to which equation (12) refers when the wavelength λ is presumed to be infinitesimal and the curvature of the surface of the imagined body is nowhere supposed to become infinitely large.

Equation (12) represents ϕ_o (i.e., the value of ϕ for an arbitrary point o of the space considered) as the sum of terms originating from luminous point ι and from luminous points lying on the boundary surface of that space. Assume point o to be infinitely close to this boundary surface, to be precise, so close that its distance from the surface is infinitesimal even against λ. The waves of light that impinge on it can then be described partly as incident, and partly as reflective or refractive, depending on whether they are moving toward the boundary or moving away from it. The luminous points from which the first originates are those situated on **one** side of the infinite plane, the luminous points whence the latter originate are those situated on the **other** side of it, which lies parallel to the closest element of the boundary surface, running through point o. If, as should be assumed, there are no incident waves in the second medium, then only incident and reflected ones exist in the first medium; let ϕ_e refer to the incident waves, ϕ_r to the reflected ones, ϕ to the entire motion at that point, which is here called point o, such that

$$\phi = \phi_e + \phi_r \quad \text{and} \quad \frac{\partial \phi}{\partial N} = \frac{\partial \phi_e}{\partial N} + \frac{\partial \phi_r}{\partial N}.$$

Here then equations (30) apply, if the incident light consists of only one wave system, and corresponding ones indicated there, if more incident wave systems must be distinguished.

One particularly simple case that is easier to imagine than the general one is having a **black** body as the second medium, i.e., one that neither reflects light nor allows it to pass through. A body through which light has the same propagation velocity as through its transparent environs and is sufficiently strongly absorbed must, according to experience, possess this property. On the surface of such a body, as assumed above, no incident waves exist, which is true of any opaque body; moreover, the quantity denoted as c is always equal to zero in this case; the condition to be satisfied at the black body's surface is therefore

$$\phi_r = 0 \quad \text{and} \quad \frac{\partial \phi_r}{\partial N} = 0. \tag{31}$$

If the body conceived for equation (12) is black and its surface is uniformly convex, then accordingly the values of ϕ and $\frac{\partial \phi}{\partial N}$ for the surface are simply found. If one imagines a plane passing by the body on a parallel tangential plane in infinitesimal proximity, then the entire surface on **one** side of this plane is situated in such a way that each element ds can offer just one contribution to ϕ_r but none to ϕ_e. The cone is imagined with its tip at luminous point ι and touching the surface; its contact line cuts the surface into two segments, one of which is turned toward the luminous point, the other away from it; at one point that is infinitely close to the first segment, luminous point ι delivers to ϕ_e the amount ϕ^*, for a point infinitesimally close to the second it delivers this amount to ϕ_r, where ϕ^* again refers to the motion that would take place if the black body were not present. Thus at the first segment

$$\phi = \phi^*, \quad \frac{\partial \phi}{\partial N} = \frac{\partial \phi^*}{\partial N}; \tag{32}$$

at the second,

$$\phi_e = 0, \quad \frac{\partial \phi_e}{\partial N} = 0,$$

and from this, pursuant to (31),

$$\phi = 0, \quad \frac{\partial \phi}{\partial N} = 0. \tag{33}$$

For a given shape of the black body, condition (31) is satisfied by determining equations (32) for the points at which straight lines from point ι meet the surface for the **first time** and equations (33) for all the other points of the surface. Under this assumption there follows from a statement proved in §3 that the integral $\int ds\, \Omega$, extended over the whole surface, vanishes when point o is chosen to be infinitesimally close to the **first** segment and that it is $= -4\pi\phi_o^*$, if point o is chosen to be infinitesimally close to the **second** segment of the surface; from which, by means of equation (12) equations (31) result for the entire surface.

From the statement just applied there follows, furthermore, that where point o is also assumed in the transparent medium, $\phi_o = \phi_o^*$, provided the straight line connecting ι and o does not meet the body surface, and $\phi_o = 0$, provided this line intersects the surface twice or more times. Because one may interpret ϕ as any one of the displacements u, v, w, it is thus explicit that in the first of the two distinguished cases the motion of light at point o is the same as if the black body were absent; in the second, however, darkness occurs at the location of o; thus it is stated that the black body casts a **shadow**, [and] that the light from the luminous point propagates **rectilinearly** in **rays** that can be regarded as independent of one another.

§5

The theorem just used, stated at the beginning of §3, is valid only under certain conditions indicated there; if they are not fulfilled then the conclusions drawn here from the theorem are not right either; **diffraction phenomena** then appear.

Imagine the luminous point ι surrounded by a black screen in which there is an aperture. Call the line, along which a cone with its tip at point ι touches the screen's surface, the **rim** of the aperture; it divides the screen surface into an inner and an outer segment. Let a given surface bounded by the rim and forming a closed surface with the one and the other of these segments surrounding the luminous point be surface s. If point o lies somewhere outside of these closed surfaces, then according to equation (12), following the hypothesis posed with reference to black bodies, therefore pursuant to equations (32), (33) and according to equation (10),

$$4\pi\phi_o = \int ds\, \Omega, \tag{34}$$

where in establishing Ω, ϕ^* should be substituted for ϕ and the integration extended over surface s. Diffraction phenomena can appear in the proximity of point o if, for a finite segment of surface s or its boundary, $r_\iota + r_o$ is constant up to an infinitesimal value, or if the straight line connecting the points ι and o passes infinitesimally close by the boundary of surface s. For the phenomena that Fresnel observed at the axis of a circular aperture or a circular screen with a luminous point on that axis, r_ι and r_o — hence also $r_\iota + r_o$ — approach constancy for all boundary points of s; for the diffraction phenomena named after Fresnel, specifically the fringes appearing close to a screen's shadow boundary, the connecting line between ι and o runs close by this boundary; for Fraunhofer diffraction phenomena (if they are represented without the use of lenses, thus on an infinitely faraway plate by means of an infinitely faraway luminous point), $r_\iota + r_o$ are virtually constant for the entire aperture.

In order to find the light intensity at point o also in these cases, first set, in accordance with equation (3),

$$\phi^* = \frac{1}{r_\iota} \cos\left(\frac{r_\iota}{\lambda} - \frac{t}{T}\right) 2\pi. \tag{35}$$

Ω than acquires the value indicated in (13). Both terms of which it is composed are of dissimilar order of quantity, since λ is infinitesimal, unless

$$\frac{\partial r_\iota}{\partial N} - \frac{\partial r_o}{\partial N}$$

is infinitesimal, which case does not need to be taken into account here. Equation (34) therefore yields

$$\phi_o = \frac{1}{2\lambda} \int \frac{ds}{r_\iota r_o} \left(\frac{\partial r_\iota}{\partial N} - \frac{\partial r_o}{\partial N}\right) \sin\left(\frac{r_\iota + r_o}{\lambda} - \frac{t}{T}\right) 2\pi.$$

To avoid lengthiness, now assume that surface s is a plane, that its dimensions compared to r_ι and r_o are so small that r_ι and r_o can be regarded as constant where they occur beyond where the sine symbol appears as well as its derivative taken for N, and finally, that the lines r_o form infinitesimal angles with the extensions of the lines r_ι. One then has

$$\frac{\partial r_o}{\partial N} = -\frac{\partial r_\iota}{\partial N}$$

and

$$\phi_o = \frac{1}{\lambda r_\iota r_o} \frac{\partial r_\iota}{\partial N} \int ds \sin\left(\frac{r_\iota + r_o}{\lambda} - \frac{t}{T}\right) 2\pi.$$

Now generalize the expression for ϕ^* in the way that equation (4) is derived from equation (3) so that you get

$$\phi^* = \frac{D}{r_\iota} \cos\left(\frac{r_\iota}{\lambda} - \frac{t}{T}\right) 2\pi + \frac{D'}{r_\iota} \sin\left(\frac{r_\iota}{\lambda} - \frac{t}{T}\right) 2\pi, \tag{36}$$

where D and D' depend on the orientation of the ray going from luminous point ι through point (x, y, z). Thereby, then,

$$\phi_o = \frac{1}{\lambda r_\iota r_o} \frac{\partial r_\iota}{\partial N} \left\{ D \int ds \sin\left(\frac{r_\iota + r_o}{\lambda} - \frac{t}{T}\right) 2\pi - D' \right.$$

$$\left. \times \int ds \cos\left(\frac{r_\iota + r_o}{\lambda} - \frac{t}{T}\right) 2\pi \right\},$$

where D and D' have the same meaning. One may now interpret ϕ as any one of the displacements u, v, w; doing this and writing A and A', B and B', C and C' for D and D', depending on whether ϕ is set $= u, v, w$, then with the unit for the light intensity defined in §1, the intensity of the light in the diffracting aperture becomes

$$= \frac{1}{2r_\iota^2}(A^2 + A'^2 + B^2 + B'^2 + C^2 + C'^2).$$

If one denotes this as J and sets

$$c = \int ds \cos \frac{r_\iota + r_o}{\lambda} 2\pi$$

$$s = \int ds \sin \frac{r_\iota + r_o}{\lambda} 2\pi,$$

then the intensity at point o becomes

$$= J\frac{1}{\lambda^2 r_o^2}\left(\frac{\partial r_\iota}{\partial N}\right)^2 (c^2 + s^2),$$

which equation numerous measurements have demonstrated to agree with experience.[5]

§6

The just-derived equation essentially assumes that the dimensions of the diffraction aperture are very large compared to the wavelengths and it is not legitimate to apply it to **diffraction spectra** which are often produced with gratings with slits just a few wavelengths in breadth.[6] Nevertheless the measurements to which we owe our knowledge about wavelengths have shown that this application yields the correct **locations** of the light maximums with great accuracy. This fact is also explained by the hypotheses grounded here by the following considerations.

Consider the grating, for which no special assumption need be made about its properties, e.g., it could just as well be a wire grating or a soot grating or a diamond grating, closely fitted inside the aperture of a planar black screen that extends on all sides to infinity. ds is understood as one element of the grating's plane or, to speak more precisely, one element of

---

[5][Orig. note 1, p. 664:] Comp. Fröhlich, Wiedemann's *Annalen*, vol. 6, p. 429 [see footnote 1 above].

[6][Orig. note 2, p. 664:] Comp. Fröhlich, Wiedemann's *Annalen*, vol. 6, p. 430 and vol. 15, p. 592 [see footnote 1 above].

a plane that lies very close to that of the grating, on the side on which point o is to be found. Equation (9) then holds; and this is simplified by introducing the assumption that r_o is infinitely large, in

$$4\pi\phi_o(t) = -\int \frac{ds}{r_o}\left\{f\left(t - \frac{r_o}{a}\right) + \frac{1}{a}\frac{\partial r_o}{\partial N}\frac{\partial\phi\left(t - \frac{r_o}{a}\right)}{\partial t}\right\}.$$

Let the plane whose element is called ds be the xy-plane of the system of co-ordinates, the x-axis vertically on the grooves, the origin the center of the grating assumed to be rectangular; furthermore, let ρ_o be the length of the line drawn from the origin to point o and let $\alpha_o, \beta_o, \gamma_o$ be the cosines of the angles they form with the coordinate axes. One then has

$$r_o = \rho_o - \alpha_o x - \beta_o y, \quad \frac{\partial r_o}{\partial N} = \gamma_o$$

and

$$ds = dx\,dy.$$

Furthermore,

$$\phi(t) = A\cos\frac{t}{T}2\pi + A'\sin\frac{t}{T}2\pi$$

$$f(t) = \frac{\partial\phi(t)}{\partial N} = B\cos\frac{t}{T}2\pi + B'\sin\frac{t}{T}2\pi$$

$$\frac{1}{a}\frac{\partial\phi(t)}{\partial t} = \frac{2\pi}{\lambda}A'\cos\frac{t}{T}2\pi - \frac{2\pi}{\lambda}A\sin\frac{t}{T}2\pi,$$

where A, A', B, B' are functions of x and y. Substituting these expressions in the equation posed for ϕ_o, one obtains through appropriate shifting of the origin for time

$$\phi_o = \iint dx\,dy\left\{C\cos\left(\frac{t}{T} + \frac{\alpha_o x + \beta_o y}{\lambda}\right)2\pi\right.$$

$$\left. + C'\sin\left(\frac{t}{T} + \frac{\alpha_o x + \beta_o y}{\lambda}\right)2\pi\right\},$$

where C and C' are inversely proportional to ρ_o, linear functions of γ_o and — what should here be stressed — linear homogeneous functions of A, A', B, B', whose coefficients do not depend on x and y. Now let the light source be a luminous point lying on the negative z-axis in infinity, $2b$ be the length of the grooves, $2n$ the number of them, and e the distance between the relevant successive pairs of points, hence $2ne$ is the grating's breadth.

It is then permissible to assume that A, A', B, B' and therefore also C and C' depend on y in such a way that they stay constant if y varies from $-b$ to $+b$, and vanish if y lies outside of this interval; on x, however, they depend in such a way that they are periodic by e if x has a value between $-ne$ and $+ne$ and vanish for other values of x. As a consequence of this, initially

$$\phi_o = \frac{\sin \frac{\beta_o b}{\lambda} 2\pi}{\frac{\beta_o}{\lambda}\pi} \int_{-ne}^{ne} dx \left\{ C \cos \left(\frac{t}{T} + \frac{\alpha_o x}{\lambda} \right) 2\pi + C' \sin \left(\frac{t}{T} + \frac{\alpha_o x}{\lambda} \right) 2\pi \right\}.$$

Since λ can be regarded as infinitesimal compared to b, the factor preceding the integral symbol is infinitesimal compared to b for any finite value of β_o, whereas it is finite if β_o is of the order of quantity $\frac{\lambda}{b}$. Conceive the integral symbol as C and C' extended over the sine and cosine of the multiple of $\frac{x}{e}2\pi$; the integrals

$$\int_{-ne}^{ne} dx \cos h\frac{x}{e}2\pi \sin \alpha_o \frac{x}{\lambda}2\pi \quad \text{and} \quad \int_{-ne}^{ne} dx \sin h\frac{x}{e}2\pi \cos \alpha_o \frac{x}{\lambda}2\pi$$

then emerge if h is defined as an integer or zero, which vanish, and the integrals

$$\int_{-ne}^{ne} dx \cos h\frac{x}{e}2\pi \cos \alpha_o \frac{x}{\lambda}2\pi \quad \text{and} \quad \int_{-ne}^{ne} dx \sin h\frac{x}{e}2\pi \sin \alpha_o \frac{x}{\lambda}2\pi,$$

which are

$$= \frac{\sin ne2\pi \left(\frac{h}{e} - \frac{\alpha_o}{\lambda} \right)}{2\pi \left(\frac{h}{e} - \frac{\alpha_o}{\lambda} \right)} + \frac{\sin ne2\pi \left(\frac{h}{e} + \frac{\alpha_o}{\lambda} \right)}{2\pi \left(\frac{h}{e} + \frac{\alpha_o}{\lambda} \right)}$$

and

$$= \frac{\sin ne2\pi \left(\frac{h}{e} - \frac{\alpha_o}{\lambda} \right)}{2\pi \left(\frac{h}{e} - \frac{\alpha_o}{\lambda} \right)} - \frac{\sin ne2\pi \left(\frac{h}{e} + \frac{\alpha_o}{\lambda} \right)}{2\pi \left(\frac{h}{e} + \frac{\alpha_o}{\lambda} \right)}$$

respectively. These expressions are infinitesimal compared to ne if λ is described as infinitesimal compared to ne; however, they are finite provided

$$\alpha_o \pm h\frac{\lambda}{e}$$

is of the order of $\frac{\lambda}{ne}$.

Because ϕ may be interpreted as any given displacement of u, v, w, it thus follows that for

$$\alpha_o = \pm h\frac{\lambda}{e}, \quad \beta_o = 0$$

the light intensity is infinitely large compared to the occurrences at all the other points within our scope; and this is what observations have shown.

$$§7$$

From these completed arguments it is easy to derive the law of **reflection** of rays of light. Place an arbitrary body opposite luminous point ι. To simplify this case, however, imagine the surface of this body enveloped in a black shell in which there is only one tiny opening on the side turned toward the luminous point; furthermore let the geometric relations be such that the reflected beam that, based on experience, forms does not hit the body's surface a second time. Again let the symbol ϕ^* refer to the motion that would occur if there were no foreign body present and let ϕ^* initially be defined by equation (35). The conditions to be fulfilled are then satisfied by setting:

For the free part of the surface:

$$\phi_e = \phi^*, \qquad \frac{\partial \phi_e}{\partial N} = \frac{\partial \phi^*}{\partial N},$$

hence, according to (30)

$$\phi_r = \frac{c}{r_\iota} \cos\left(\frac{r_\iota}{\lambda} - \frac{t+\gamma}{T}\right) 2\pi, \qquad \frac{\partial \phi_r}{\partial N} = -c \frac{\partial}{\partial N} \frac{1}{r_\iota} \cos\left(\frac{r_\iota}{\lambda} - \frac{t+\gamma}{T}\right) 2\pi,$$

and consequently

$$\phi = \phi^* + \frac{c}{r_\iota} \cos\left(\frac{r_\iota}{\lambda} - \frac{t+\gamma}{T}\right) 2\pi$$

$$\frac{\partial \phi}{\partial N} = \frac{\partial \phi^*}{\partial N} - c \frac{\partial}{\partial N} \frac{1}{r_\iota} \cos\left(\frac{r_\iota}{\lambda} - \frac{t+\gamma}{T}\right) 2\pi,$$

For **those** points of the blackened part of the surface hit for the first time by the line extending away from luminous point ι:

$$\phi = \phi^*, \qquad \frac{\partial \phi}{\partial N} = \frac{\partial \phi^*}{\partial N},$$

For all the other points of the blackened surface:

$$\phi = 0, \qquad \frac{\partial \phi}{\partial N} = 0.$$

Pursuant to equations (12) and (11), the excess of the value for ϕ_o above the value that ϕ_o would have if the **entire** surface of the foreign body were blackened would then be the sum of the two integrals

$$-\frac{1}{4\pi} \int c \frac{ds}{r_\iota r_o} \left(\frac{1}{r_\iota} \frac{\partial r_\iota}{\partial N} + \frac{1}{r_o} \frac{\partial r_o}{\partial N} \right) \cos \left(\frac{r_\iota + r_o}{\lambda} - \frac{t + \gamma}{T} \right) 2\pi$$

and

$$\frac{1}{2\lambda} \int c \frac{ds}{r_\iota r_o} \left(\frac{\partial r_\iota}{\partial N} + \frac{\partial r_o}{\partial N} \right) \sin \left(\frac{r_\iota + r_o}{\lambda} - \frac{t + \gamma}{T} \right) 2\pi, \qquad (37)$$

where the integral is extended over the open part of the surface — that may be called s.[7] The first of these two integrals is negligible against the second if point o is located at a finite distance from the surface, because λ is infinitesimal, so the mentioned difference between the two values for ϕ_o is described by integral (37).

This also holds if ϕ^* is given by equation (36) instead of by equation (35); just the values for c and γ are then different. Integral (37) has the form of integral (19); from the considerations performed with reference to the latter there follows that the former generally vanishes. Integral (19) does not vanish if surface s is intersected by the line connecting points ι and o, but integral (37) vanishes even then because then

$$\frac{\partial r_\iota}{\partial N} + \frac{\partial r_o}{\partial N} = 0$$

for the intersection point. Integral (37) differs from zero if there is a point on surface s whose connecting line with points ι and o forms the same angles with the normal of surface s and lies on the same plane. Thus is expressed that reflected rays exist and what orientations they have. A disturbance by diffraction phenomena occurs if on one finite part of surface s or its boundary $r_\iota + r_o$ is constant up to an infinitesimal value or if point o lies infinitely close to the boundary of the reflected beam.

From these theorems determining the orientations of reflected rays just derived, it is possible to develop the geometrical properties of a beam emitted from a luminous point and reflected off a curved surface. The

---

[7][Orig. note 1, p. 668:] It may be demonstrated without difficulty that if point o lies on the surface or is infinitesimally close to it, this expression leads back to the assumed values for ϕ and $\frac{\partial \phi}{\partial N}$. However, this proof will not be presented here.

computations carried out in §3 also allow one to indicate how the intensity of a ray from such a beam varies as well as the phase from one point to another.

The part of ϕ_o corresponding to the reflected light, i.e., the expression (37), is given by the expressions (24), (25) or (26), if therein

$$G = \frac{K}{\rho_o},$$

where K means a quantity independent of ρ_o. From this follows that the intensity of a reflected ray varies with ρ_o such that it is inversely proportional to the absolute value of

$$\rho_o^2 \mu_1 \mu_2.$$

According to equations (27) and (22) this expression may be written as

$$(b_{11}\rho_o + c_{11})(b_{22}\rho_o + c_{22}) - (b_{12}\rho_o + c_{12})^2,$$

where the quantities b and c are independent of ρ_o and

$$c_{11} = \frac{1}{2}(1 - \alpha_o^2), \quad c_{12} = -\frac{1}{2}\alpha_o\beta_o, \quad c_{22} = \frac{1}{2}(1 - \beta_o^2).$$

If $\rho_o = f_1$ and $\rho_o = f_2$ are (always real) roots of the quadratic equation that one obtains by setting this expression equal to zero, then the intensity thus also is inversely proportional to the absolute value of

$$(\rho_o - f_1)(\rho_o - f_2).$$

At points $\rho_o = f_1$ and $\rho_o = f_2$ the intensity is infinite; these are the ray's foci.

With regard to the phase it should be noted that, as expressions (24), (25), (26) show, they vary jump-wise by $\frac{\pi}{2}$ if point o runs through one of the foci.

It scarcely needs mentioning that entirely similar considerations to those about reflection can be made about the refraction of light rays.

Kirchhoff's Theory for Optical Diffraction, Its Predecessor and Subsequent Development: The Resilience of an Inconsistent Theory, by Jed Z. Buchwald (Caltech) and Chen-Pang Yeang (Univ. of Toronto)

Abstract

Kirchhoff's 1882 theory of optical diffraction forms the centerpiece in the long-term development of wave optics, one that commenced in the 1820s when Fresnel produced an empirically successful theory based on a reinterpretation of Huygens' principle, but without working from a wave equation. Then, in 1856, Stokes demonstrated that the principle was derivable from such an equation albeit without consideration of boundary conditions. Kirchhoff's work a quarter century later marked a crucial, and widely influential, point for he produced Fresnel's results by means of Green's theorem and function under specific boundary conditions. In the late 1880s, Poincaré uncovered an inconsistency between Kirchhoff's conditions and his solution, one that seemed to imply that waves should not exist at all. Researchers nevertheless continued to use Kirchhoff's theory — even though Rayleigh, and much later Sommerfeld, developed a different and mathematically consistent formulation, which, however, did not match experimental data better than Kirchhoff's theory. After all, Kirchhoff's formula worked quite well in a specific approximation regime. Finally, in 1966 Marchand and Wolf employed the transformation of Kirchhoff's surface integral that had been developed by Maggi and Rubinowicz for other purposes. The result yielded a consistent boundary condition that, while introducing a species of discontinuity, nevertheless rescued the essential structure of Kirchhoff's original formulation from Poincaré's paradox.

1. Introduction

On 22 June 1882, the University of Berlin's professor of theoretical physics, Gustav Robert Kirchhoff (1824–1887), read an influential paper titled "Zur Theorie der Lichtstrahlen" ("On the theory of light rays") to a meeting of the Prussian Academy of Sciences in Berlin. The purpose of the paper was to deduce from the wave equation the expression governing the diffraction of light by an aperture on an otherwise opaque screen. To do so Kirchhoff assumed a particular set of boundary conditions: namely, that both the amplitude of the disturbance as well as its spatial gradient vanished on the screen, but that they remained unaltered over the aperture itself.[1] He was in this way able to generate a solution for scalar diffraction that could yield the empirically-successful (so far as was then known) expression that Augustin Fresnel (1788–1827) had produced six decades before using an altogether different line of argument (on which more below). Despite its frequent presence in physicists' and engineers' publications, Kirchhoff's theory of diffraction has not until recently attracted the attention of physics (or mathematics) historians for understandable reasons: historical focus has principally aimed at episodes in 19th century optics and electromagnetism that brought either fundamental changes — e.g. the wave theory of light, electromagnetic field theory, kinetic theory and statistical mechanics — or influential technological breakthroughs — e.g. Hertz's production of electric waves and the subsequent invention of wireless telegraphy. Kirchhoff's theory fits neither criterion. It did not introduce any novel physical entities or mechanisms beyond what wave optics had stipulated; nor did it lead to technological innovation. It nicely fits, one might say, Thomas Kuhn's conception of "normal science", in which practitioners solve problems that arise within a given system without violating its fundamental boundaries.[2]

   Although Kirchhoff's theory did not have significant ontological or technological implications, it nonetheless raised important questions concerning the use of mathematics in theoretical physics. What made the theory interesting in subsequent years is the mathematical inconsistency of the boundary conditions that were used. So far as was known at the time, however, his solution worked quite well empirically, and Kirchhoff himself never

---

[1] Kirchhoff 1882, also printed as Kirchhoff 1883. In this article, we use an English translation by Ann Hentschel 2016, here to be found on pp. 31–61 of this anthology.

[2] Kuhn 1962, pp. 10–34.

remarked the inconsistency. Decades after the French mathematician Henri Poincaré (1854–1912) published a deleterious consequence of the inconsistency in 1892,[3] the theory nevertheless continued to appear in major textbooks and research periodicals in optics and electromagnetism. Physicists and engineers treated it not as an antiquated and inconsistent effort to derive empirically workable results, but as a good enough working model, for Kirchhoff's solution nicely fit both optical and microwave experimental data under particular, but commonly applicable, conditions. Indeed, physicists' and engineers' interest in, and use of, Kirchhoff's theory has certainly not waned over the decades.[4]

The persistent deployment of an apparently inconsistent theory even after recognition of its flaws is not unique in the history of physics. The infamous divergence of the quantum field integrals in self-energy calculations in quantum electrodynamics (QED) and physicists' various ad-hoc manipulations to bypass the problem before the introduction of renormalization provides one noteworthy example.[5] Another concerns early 20th-century Cambridge mathematicians' continued use of circulatory theory to explain airfoil lifting despite a salient contradiction — namely, d'Alembert's paradox, according to which there should be no lift at all in a perfect fluid — that had been well known for centuries.[6] In these cases, a major reason for the tenacity of an admittedly problematic theory was essentially pragmatic: the physical-mathematical problem was simply too complex, and no comparable alternative was available at the time.

Despite the similarities, Kirchhoff's account of diffraction differed in one essential respect from these two examples that makes this situation particularly compelling. In the first decades of the twentieth century, Lord Rayleigh (1842–1919) and Arnold Sommerfeld (1868–1951) derived different solutions to the same diffraction problem under a set of consistent boundary conditions. Unlike the cases of QED before renormalization and the circulatory theory of airfoil lifting in the early twentieth century, therefore, a mathematically consistent alternative was in fact available. Yet the existence of a seemingly more appealing and logical alternative

---

[3] Poincaré 1892b, pp. 187–188.

[4] According to Werner Marx's bibliometric analysis in this volume, Kirchhoff's 1882 work from 1900 to 2010 has been specifically cited 70 times in research articles; moreover, the citation count has increased over the years.

[5] Schweber 1994.

[6] Bloor 2011.

did not eliminate or marginalize Kirchhoff's theory. This mathematically inconsistent (and physically untenable) solution has continued to appear and thrive in textbooks and periodicals as the standard approach to the problem of diffraction. Why, one may ask, did scientists continue to stay with Kirchhoff's theory despite the presence of a consistent alternative?

Recently, several historians and philosophers have begun to pay closer attention to this curious episode.[7] These studies deepen our understanding of conceptual and technical aspects of Kirchhoff's theory and help us clarify its philosophical implications, but in what follows we are concerned with the details of the theory's persistence and with the consistent alternatives to it. The theory's empirical success within the experimental regime of 19th and early 20th century optics certainly goes reasonably far in accounting for its persistence. More, however, was involved than the results of experiment. The theory's tenacity also reflects the hold of a long-standing intellectual tradition in optics. Rooted in a principle introduced by Christiaan Huygens (1629–1695) in 1678 and first deployed for diffraction by Augustin Fresnel (1788–1827) in 1818, this approach calculated wave intensity by means of a single integral over regions that are not blocked by a diffracting object. Such an integral could be interpreted as comprised of waves, or wavelets, that emanate from each of the points on the object's open regions and so on the incident wave front proper. Even after the scope of the inconsistency became altogether clear, physicists continued to seek (and to find) a single integral that could give physical meaning at least to a form of Huygens' principle, if not to the original, a principle that Kirchhoff's theory so nicely exemplified. In what follows we explore the structure of Kirchhoff's theory and the alternatives to it in order to bring out the several ways in

---

[7]For a comprehensive historical study of Kirchhoff's 1882 paper by translating it into English, presenting a commentary, situating it within Kirchhoff's intellectual biography, and conducting a scientometric analysis of citations to it, cf. the contributions by Klaus Hentschel, Ning Yan Zhu, Ann Hentschel, and Werner Marx in this volume. In light of the realism debate in philosophy of science, Juha Saatsi and Peter Vickers have used Kirchhoff's theory as a counterexample to primitive realism — what they have termed "naïve optimism" — which contends that any significant novel predictive success can be explained by the truth content of the assumptions that play an essential role in the derivation. Kirchhoff's boundary conditions, Saatsi and Vickers noted, are both mathematically inconsistent and physically untenable. Yet they are essential assumptions for the derivation of Kirchhoff's empirically-successful (within certain regimes) formula (J. Saatsi & P. Vickers 2011). We thank Ning Yan Zhu for catching a number of misprints and missing or incorrect references in a previous version of the present article.

which mathematical structures that exemplify principles akin to Huygens' generated theories of diffraction.

2. Fresnel Applies Huygens' Principle to Diffraction

When Fresnel first tackled the problem of diffraction neither he nor Thomas Young (1773–1829) before him had deployed Huygens' principle. Instead, both Young and Fresnel initially considered interference to take place between rays of light emanating directly from the source and from emission stimulated by the source at the edges of diffracting objects. That way of thinking essentially conformed to a long-standing tradition that only Huygens himself had challenged, one in which the central physical entity in optics was the ray of light.[8] Fresnel broke with that tradition and introduced Huygens' principle when confronted with an empirical conflict that arose in a particular situation. But to deploy the principle Fresnel had to develop a method for decomposing waves from a multiplicity of different loci into ones with common phases but different amplitudes. Such a method had emerged previously in his exploration of the chromatic effects generated by the passage of polarized light through thin, birefringent crystals.

To achieve that result Fresnel assumed a wave of the form $a \sin[2\pi(ft - x/\lambda)]$, with λ the wavelength and f the frequency of the disturbance. If the wave emanating from the source reaches a point via two different paths, one of which traverses a distance x while the other requires an additional distance δ then the latter will be $a \sin[2\pi(ft - x/\lambda) - i]$ where i is $2\pi\delta/\lambda$. Algebraically decomposing this expression, Fresnel could split the second wave into two parts:

$$a \sin[2\pi(ft - x/\lambda) - i] = a \cos(i) \sin[2\pi(ft - x/\lambda)]$$
$$- a \sin(i) \cos[2\pi(ft - x/\lambda)].$$

That is, a single wave with arbitrary phase i can always be considered to arise from two other waves with amplitudes $a \cos(i)$, $a \sin(i)$ that differ in phase by 90°. This involves the same kind of process, Fresnel noted, as the composition of two mutually-perpendicular forces of magnitudes equal to the component amplitudes. Whence, he concluded, the magnitude of the resultant of two waves with the same frequency that otherwise

---

[8]On the history of Huygens' optics and its background see Shapiro 1973. On Young see Kipnis 1991, and on Fresnel see Buchwald 1989.

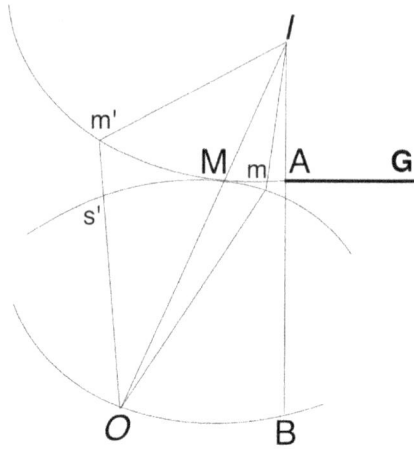

Figure 5: Fresnel's configuration for diffraction.

differ in phase by 90° is just the square root of the sum of the squares of the component amplitudes. This decomposition enabled the application of Huygens' principle to diffraction.

Consider with Fresnel a point light source I, a semi-infinite plane obstacle AG whose edge A is directly beneath I, and a screen parallel to AG. His objective was to calculate the resultant wave at a point O on the screen but outside the shadow cast by AG (Figure 5). According to Huygens' principle the wave intensity at O should be the sum — or integral for an effectively continuous distribution — of all the wavelets emanating from the front $m'MmA$, calculated as it passes the obstacle AG.

Fresnel assumed that the remainder of the wave was completely blocked by the obstacle, that the obstacle neither modified the wave's spherical shape, nor affected the phase of the vibrations at any unblocked point of the front. Note that in so assuming Fresnel essentially ignored the finite dimensions of such an obstacle in that he did not separately consider what the wave front might be on the side of the obstacle facing the luminous point and on the side facing the screen point. He simply annulled the wave altogether at the obstacle's locus. We shall see below that the introduction of a solution to the scalar wave equation by Kirchhoff mandated explicit consideration of both surfaces.

The resultant at O was, then, the integral of all the wavelets over the semi-infinite front $m'MmA$ whose phases were respectively proportional to the path lengths AO, mO, MO, $m'O$, etc. To carry out this integral,

Fresnel had to choose a convenient origin. For various reason he selected M (subsequently termed the 'pole'), the intersection between the direct line IO and the wave front $m'MmA$.[9] Fresnel easily calculated that the distances m's' (s' is the intersection between the line $m'O$ and the arc around the center O) by which the wavelets differ in their distances to the point of observation are to a very good approximation for points reasonably near the pole $z^2(b+c)/(2bc)$ with b representing IA and c representing AB, while z is the distance Am'. Consequently the phase difference i in Fresnel's original decomposition becomes $\pi z^2(b+c)/bc\lambda$. Each of the wavelets decomposes accordingly, with the consequence that the square of the resultant amplitude (and so the optical intensity) at the observation point must be:

$$\left[\int dz \sin(\pi z^2(b+c)/bc\lambda)\right]^2 + \left[\int dz \cos(\pi z^2(b+c)/bc\lambda)\right]^2.$$

Fresnel's successful demonstration that his formulae worked well empirically won him a Paris Academy prize in 1818 despite the presence on the judging committee of no less than Pierre-Simon Laplace (1749–1827) and Siméon Denis Poisson (1781–1840), neither of whom was sympathetic to wave optics. Today, his expression is a particular approximation to wave propagation under a specific condition: it is the second-order expansion of the propagating phase that takes into account the spherical wave front when the distance of interest lies between the quasi-static near-field zone and the plane-wave-like far-field zone.[10] At the time, and for decades thereafter, his expression had a broader and more profound meaning, for it demonstrated that, far from being a curious conception which could at best be used to generate known results in a hitherto-unacceptable theory, Huygens' principle had proven empirical consequences. The result was to entrench the principle as a physically-meaningful foundation over the following decades as the wave theory of light was gradually assimilated, understood and produced altogether novel results. Most of these novelties did not depend on the principle itself since they were for the most part concerned with polarization phenomena. Nevertheless, as a unifying physical concept with a specific mathematical expression Huygens' principle provided a powerful foundation for this new, complex and difficult theory. John Herschel (1792–1871), who published the first comprehensive article on wave optics in 1827, placed

---

[9] For full details see Buchwald 1989, chap. 6.
[10] See, for example, Kong 1986, pp. 671–95.

particular emphasis on the principle, which he expressed in the following
way:

> ...conceive the surface of any wave A B C to consist of vibra-
> tory molecules, *all in the same phase of their vibrations* (*sic*).
> Then will the motion of any point X ...be the same, whether it
> be regarded as arising from the original motion of S [the source],
> or as the resultant of all the motions propagated to it from all
> the points of the surface.[11]

The essence of this way of treating the problem can be concisely
expressed. The diffracted wave intensity $u(\vec{r}, t)$ at a given location $\vec{r}$ and
time t is taken to be

$$u(\vec{r}, t) = \int_\Omega dr' A(\vec{r}, \vec{r}') u_{in}(\vec{r}', t - |\vec{r} - \vec{r}'|/c).$$

The integration takes place over the open parts of a blocking screen, while
$u_{in}(\vec{r}', t)$ is the intensity of the wave directly from the light source at a point
$\vec{r}'$ at time t, $A(\vec{r}, \vec{r}')$ is the inclination and any other necessary factors
for propagation from $\vec{r}''$ to $\vec{r}$, c is the wave speed, and $|\vec{r} - \vec{r}'|/c$ is the
time delay (or, equivalently, phase factor) from $\vec{r}'$ to $\vec{r}$. In this manner of
working, which remained common for decades after Fresnel's original work,
the solution to the wave equation is presumptively given (because of the
known u_{in}) under the assumption that the screen has no other effect than
to stop propagation except at its open parts. That is, the source wave is
assumed to be completely unaltered at the screen's openings and completely
extinguished over its surface.

3. Stokes' "Dynamical Theory of Diffraction"

Neither Fresnel nor those who worked in wave optics for decades afterwards began their analysis of diffraction with a differential equation. In
other areas of optics, such as dispersion and birefringence, analysis did proceed in the 1830s by generating solutions to a wave equation, but in those
cases boundary issues were irrelevant, posing a very different type of problem. Nevertheless, the physical foundation of wave optics then required the
existence of an *ether*, a substance that pervades space and material bodies and that must be governed by the laws of mechanics. In France and

---

[11] Herschel 1827, sec. 623.

Germany for decades the ether was presumed to be constituted of particles governed by Coulomb-like forces of various presumptive intensities, whereas in Britain after *circa* 1830 it was widely treated as an effective continuum governed by variously-assumed constitutive stipulations.[12]

Fresnel himself never deployed a differential equation for optics to any major extent, though he certainly did consider what sort of forces must exist between the elements of the ether in order to justify his assumptions concerning the relationships required between wave speed and direction of oscillation in birefringent media. That is, the only circumstances at the time that raised questions of the forces that act on ether elements were ones in which the wave speed varies with the direction of propagation and of oscillation. Diffraction required nothing of the sort because the empirical circumstances were limited at the time to propagation in isotropic media. There was however one important question that did demand a consideration of the wave equation, namely the form of the inclination factor governing the amplitude of the Huygens wavelets as a function of the angle between the line to the observation point and the normal to the front which the wavelets comprise. Fresnel had argued on rather weak physical grounds that, in the circumstances he was considering, the amplitudes are independent of direction for the hemisphere tangent to the incident front in the direction of propagation, which meant that in his diffraction formulae the inclination factor was simply ignored. The first effort to determine that factor required a consideration of the wave equation, and it was accomplished in 1849 by George Gabriel Stokes (1819–1903).

A senior wrangler and Smith's prizeman at Cambridge University and eventually Lucasian Professor of Mathematics there, Stokes had worked on optics (including the aberration of light and spectroscopy) as well as fluid dynamics, a subject whose mathematical structure resembled that of optics precisely because the optical ether was presumptively governed by mechanical relations.[13] Stokes introduced his "Dynamical Theory" with the following words:

> When light is incident on a small aperture in a screen, the illu-
> mination at any point in front of the screen is determined, on
> the undulatory theory, in the following manner. The incident

---

[12] For the example of dispersion and Augustin Cauchy's (1789–1857) elaborate mathematics see Buchwald, 2012. For a broad overview of the period see Buchwald 2013 and Darrigol 2012.

[13] Stokes 1856; reprinted in Stokes 1883, pp. 243–328.

waves are conceived to be broken up on arriving at the aperture;
each element of the aperture is considered as the centre of an
elementary disturbance, which diverges spherically in all directions, with an intensity which does not vary rapidly from one
direction to another in the neighborhood of the normal to the
primary wave; and the disturbance at any point is found by taking the aggregate of the disturbances due to all the secondary
waves, the phase of vibration of each being retarded by a quantity corresponding to the distance from its centre to the point
where the disturbance is sought.[14]

This much is entirely similar to the standard assumption since Fresnel,
and Stokes did not go beyond it. Nevertheless, he penetrated very far
into the mathematical core of contemporary wave theory — farther than
anyone had since Fresnel — by deriving an entirely new result involving
the polarization of diffracted light that he immediately sought to confirm
in the laboratory. Stokes' primary purpose however was to uncover the
inclination or amplitude factor. To do so he began at once with the general differential equation of motion for an isotropic, inviscid elastic solid
that he had himself developed, and that he now applied to the optical
ether:[15]

$$\partial^2 \vec{u}/\partial t^2 = b^2 \nabla^2 \vec{u} + (a^2 - b^2)\nabla(\nabla \cdot \vec{u}),$$

where $\vec{u}$ is the displacement, and a, b are elastic constants. He then separated the equation by defining 'for shortness' δ as the negative compression
$\nabla \cdot \vec{u}$ (or 'dilatation' as he called it), and $\vec{\omega}$ as the rotation (or, again in
Stokes' terminology, the 'distortion') $(1/2)\nabla \times \vec{u}$:

$$\partial^2 \delta/\partial t^2 = a^2 \nabla^2 \delta,$$

$$\partial^2 \vec{\omega}/\partial t^2 = b^2 \nabla^2 \vec{\omega}.$$

The single equation for the compression, and the three for the components
of the rotation, all have precisely the same form, and Stokes could at once

---

[14] *Ibid.*, 243–244.
[15] Stokes 1845c. Of course neither Stokes nor anyone else, with the partial exception of
James Clerk Maxwell (1831–1879), used vector notation until the 1890s. Nevertheless
physicists and mathematicians of the period were able almost at once to produce the
component equivalents of even complex vector operations, so that modern notation does
not unduly alter their original understanding.

write down the following solution, which he obtained from Poisson[16]:

$$U = \frac{t}{4\pi} \int F(at)d\sigma + \frac{1}{4\pi} \frac{d}{dt} \left\{ t \int f(at)d\sigma \right\}. \tag{1}$$

Here U is the solution at some point P, t is the time, and $f(at)$, $F(at)$ are respectively the initial values of the function δ (or a component of $\vec{\omega}$) and of its time derivative at all positions whose distance from P is at (or bt). The integrals, which correspond to the mean values of the functions over the surfaces, are taken over a spherical surface of radius at (or bt) that surrounds the field point P. Poisson's solution has the peculiarity of representing the effect at P in terms of a time-dependent radius that is drawn from P. Instead, that is, of following a pulse as it expands outwards, with this solution we start at a given point and cut space with surfaces drawn about it until we find surfaces that pass through the regions which contain the initial disturbance. Poisson, as it were, held fixed the initial disturbance and went looking for it from the field point, and this solution (which is difficult to formulate in a rigorous manner[17]) applies to any disturbance that begins at some moment. In contrast to the manner in which the wave equation would be treated decades later, wherein the time-varying part, subjected to a Fourier expansion, is separated from the location-varying part, this solution lumps together space and time, treating the whole as an initial-value problem. Stokes then justified its extension from, in effect, a pulse to infinitely long wave trains in the following way:

> In the investigation it has been supposed that the force [distur-bance] began to act at the time 0, before which the fluid was at rest, so that $f(t) = 0$ when t is negative. But it is evident that exactly the same reasoning would have applied had the force begun to act at any past epoch, so that we are not obliged to suppose $f(t)$ equal to zero when t is negative, and we may even suppose $f(t)$ periodic, so as to have finite values from $t = -\infty$ to $t = +\infty$.[18]

---

[16] It is particularly ironic that Stokes took this from Poisson, because he used it to argue that the inclination factor varies in a fashion that Poisson himself would probably not have (Buchwald 1989, p. 192).

[17] On which see Baker and Copson 1939, pp. 12–15.

[18] Stokes was a bit disingenuous here, since not only his investigation, but also Poisson's solution, requires the limitation. Stokes' quick attempt to extend the class of allowable functions to cover those which are not temporally delimited requires a great deal more justification than this (Stokes 1883, p. 278).

To generate a formula that could be applied to diffraction, Stokes simply assumed the disturbance to be sinusoidal in form. The final result for the wave of distortion produced the following expression for its value at a field point located at a distance r from the disturbance on a surface element $d\sigma$ that forms part of an unblocked region:

$$\frac{d\sigma}{2\lambda r}(1 + \cos\theta)\sin\phi\cos\left[\frac{2\pi}{\lambda}(bt - r)\right]. \tag{2}$$

Here θ is the angle between the normal at the surface element and the line from there to the field point, while ϕ represents the angle between the direction of the oscillation and that same line: it was introduced to take account of polarization (Figure 6). This expression constitutes the first of an inclination factor, namely $1 + \cos\theta$.[19] Stokes' main interest here was to show that, were the expression integrated over a completely unblocked surface, then the original wave would be regenerated, demonstrating thereby the consistency of an analysis based on Huygens' principle. He did not

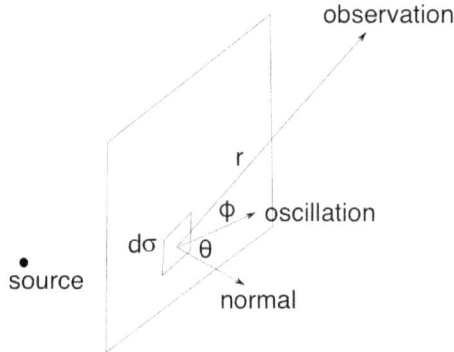

Figure 6: Stokes' configuration for diffraction.

---

[19]Stokes set the compression wave to the side. However, the mechanically-necessary existence of both compression and distortion generally posed a problem for such investigations because they cannot easily be divorced from one another, particularly if the model involves, like Cauchy's, forces between particles (on which see Buchwald 1980, 1981). Cauchy assumed the compression wave to be invisible (which raised energy issues that might in principle be detectable), while Stokes took the compression constant to be so large that the corresponding wave speed was infinitely larger than the speed of the distortional wave, implying that the former would not be visible or otherwise affect the latter. This amounted to assuming that the ether is incompressible, requiring the rate of displacement $\partial\vec{u}/\partial t$ of any element to satisfy $\nabla \cdot \partial\vec{u}/\partial t = 0$.

however go further to amend Fresnel's original formulae by incorporating the inclination factor — which would in any case be pointless given that the configuration of diffraction experiments at the time made the factor inconsequential.

Unlike Fresnel, Stokes did not have to simply assume Huygens' principle because its wavelets now appeared as the integrands of a solution. Still, Stokes had not gone any further than had Fresnel in respect to the conditions that should obtain over a blocking surface and within its open regions. He had also simply assumed that the wave function must vanish on the surface and remain completely unaffected over its open regions. The consistency that he proved concerned only the question of whether the expression he had generated for the wavelets could reproduce the original disturbance, not whether the presumed values of the wave within the bounded regions and over the blocking surface could themselves be justified.

Questions concerning the boundary conditions and uniqueness requirements that arise when working with partial differential equations had previously been dealt with in respect to the Laplace equation for the potential function ($\nabla^2 \varphi = 0$), most notably in England by the mathematician George Green (1793–1841) in 1828.[20] Green developed, among other results, what became known as "Green's theorem" (of which more below), and he did argue that, given the value of the potential over a closed surface, then there is one and only one unique solution to the Laplace or Poisson equation within or without the closed surface, provided that the potential vanishes at an infinite distance from it, or what would eventually be termed the Dirichlet principle by Bernhard Riemann (1826–1866).[21]

Now in 1845 and 1846 Stokes published two influential papers concerning the aberration of light that invoked a condition for the ether's velocity with respect to the Earth.[22] He was almost certainly not familiar with the work in which Green had deduced his theorem since it had not been published in a journal, and since in any case its title seemed to concern itself with electricity. Only in 1845, when William Thomson (1824–1907, later Lord Kelvin) rediscovered Green's *Essay*, did it become broadly known. Even if Stokes had known Green's theorem, he would not have thought it relevant to the problem of aberration, despite the fact that both Green's

---

[20] Green 1828.
[21] After Peter Lejeune Dirichlet (1805–1859); *cf.* Riemann 1857, p. 17.
[22] Stokes 1845a,b.

Essay and Stokes' aberration problem concerned the question of solutions to the Laplace equation under specific boundary conditions.

The question Stokes addressed was to find an appropriate condition on the motion of the ether near the Earth's surface that could yield the constant of aberration, i.e., the expression for the deviation of a star's apparent position as a function of the speed of light through otherwise stationary ether and the Earth's velocity through the medium. Stokes demonstrated that the appropriate expression would follow provided that two conditions were satisfied: first, that the ether's velocity at the Earth had to be the same as the Earth's with respect to the ether at a distance, i.e. that the Earth must fully drag the ether along, and second that the part of the ether's velocity due to the Earth's motion must be irrotational, i.e. that $\nabla \times \vec{v}$ must vanish, in which case that part of the velocity must be the gradient of a scalar function.[23] Since Stokes had also required the divergence of the velocity to vanish in his diffraction theory in order to eliminate the compression wave, this would have meant that the ether's velocity had to satisfy a Laplace equation were it not that Stokes had limited the requirement to a 'part' of the velocity, which accordingly essentially avoided having to deal with a boundary-value problem. He was able in this way to ignore altogether the Earth's surface by simply assuming in his calculation of the aberration constant that the small additional 'part' of the ether's velocity due to the ether's being entangled with the moving Earth was that of the Earth itself relative to an ether undisturbed except by the ($\nabla \times \vec{v}$ bearing) motions of light. Nearly half a century later, in 1887, the Dutch physicist H. A. Lorentz returned to the question and decided that Stokes' analysis had to be treated as a boundary-value problem. Whatever the mechanism of Earth-ether entanglement might entail, Lorentz effectively assumed, since the ether is acknowledged to be incompressible, the Laplace equation must be satisfied, in which case the ether's motion is specified altogether by the scalar function's normal gradient at the Earth's surface, i.e. by the velocity, assumed to be that of the Earth itself. Lorentz then demonstrated that under these circumstances the component of the gradient of the requisite scalar function tangent to the surface will differ from the corresponding component of the Earth's velocity, so that the

---

[23] Of course optical polarization required the existence of transverse oscillations, i.e. of $\nabla \times \vec{v}$, and so Stokes emphasized that irrotationality could hold only for "that part of the motion of the ether which is due to the motion of translation of the earth and planets" (*ibid.*, p. 137).

ether will slip over the surface, ruining the general applicability of Stokes' calculation.[24]

Stokes in the mid-1840s had paid no attention at all to boundary-value questions, while Lorentz in 1887 insisted on invoking one immediately by jettisoning Stokes' admittedly-vague separation of Earth-plus-ether velocity for that of the ether. Lorentz in other words constrained the problem to be one that simply had to satisfy a specific boundary condition on the Laplace equation, namely the so-called Neumann condition, according to which specification of the scalar function's normal gradient over a closed surface completely determines the function's value.[25] However, as late as the 1870s physicists at least were not generally paying close attention to the mathematical requirements imposed on harmonic functions by boundary-value requirements, as we shall now see in detail in the case of Kirchhoff.

4.   Kirchhoff Renovates Huygens' Principle

Gustav Robert Kirchhoff was well-known when he left his two-decade-long academic base in Heidelberg and took the first chair for an *Ordinarius* in theoretical physics at the University of Berlin in 1875. His Berlin years from 1875 to his death in 1887 were capstones to a career, since his major contributions — electric circuit theory, spectroscopy, and thermal radiation — had all been made well before, while at Berlin he devoted himself principally to teaching.[26] Precisely because of the need to prepare lectures Kirchhoff looked into the issues in optical diffraction that Stokes had only partially resolved. In 1882, Kirchhoff read a seminal paper on optics at the Royal Prussian Academy of Sciences in Berlin. It was published in the Academy's *Sitzungsberichte* and the following year in the *Annalen der Physik und Chemie*.

The aim of Kirchhoff's paper was similar to Stokes' three-decades earlier work but more general: Kirchhoff aimed to develop a "fully satisfactory theory" of light, not solely of diffraction *per se*, by starting

---

[24]Lorentz 1887.

[25]Named after the German mathematician Carl Neumann (1832–1925). For histories of the boundary-value criteria for harmonic functions see Cross 1985, Cheng and Cheng 2005. See Kline 1972, chap. 28 for an overview of partial differential equations in the period.

[26]Jungnickel & McCormmach 1986, pp. 30–32, vol. 2.

with the wave equation itself.[27] Like Stokes, Kirchhoff assumed that the ether displacement $\vec{u}$ corresponding to light was altogether transverse, setting aside the possible existence of a compression wave (on the widely accepted premise of an incompressible ether), yielding thereby the following equation:

$$\frac{\partial^2 \vec{u}}{\partial t^2} = a^2 \nabla^2 \vec{u}.$$

Also like Stokes, Kirchhoff used a scalar variable φ to express any of $\vec{u}$'s components, but unlike Stokes he altogether ignored polarization (which Stokes had partially taken into account through the introduction of the variable ϕ – *cf* 2):

$$\frac{\partial^2 \varphi}{\partial t^2} = a^2 \nabla^2 \varphi.$$

Kirchhoff's method of solving the equation however differed in fundamental ways from Stokes'. Stokes had treated the situation entirely as an *initial-value problem*, and he had accordingly used Poisson's formulation to express the solution in terms of the subsequent evolution of the distributed sources at time zero (*cf* 1). That solution tacitly assumed that the disturbance vanishes on a blocking surface and is otherwise unaffected, but Stokes did not directly incorporate the conditions into the wave equation's integral solution. By contrast, Kirchhoff explicitly involved the *boundaries* of the illuminated region.

He began with a point source in unbounded space. Assuming a trigonometric oscillation with period T for an outward-propagating spherical wave, Kirchhoff first assumed the following expression for the wave function:

$$\varphi = \frac{D}{r_i} \cos\left[2\pi\left(\frac{r_i}{\lambda} - \frac{t}{T}\right)\right] + \frac{D'}{r_i} \sin\left[2\pi\left(\frac{r_i}{\lambda} - \frac{t}{T}\right)\right],$$

or (following the decomposition that Fresnel had originally developed)

$$\varphi = \sqrt{\left(\frac{D}{r_i}\right)^2 + \left(\frac{D'}{r_i}\right)^2} \sin\left\{\left[2\pi\left(\frac{r_i}{\lambda} - \frac{t}{T}\right)\right] - i\right\},$$

---

[27]Kirchhoff 1883, *cf* p. 663. The paper was included in the posthumous publication of Kirchhoff's works: Kirchhoff 1891a, pp. 22–54.

in which r_i is the distance between the source or, in Kirchhoff's words, 'luminous' point and the locus of observation, while D and D' are amplitude coefficients whose ratio determines the wave's phase.[28]

Kirchhoff next set his "luminous point" to the side, together with his presumptive expressions for the wave, and turned instead to "Green's theorem" specifically in order to obtain an expression that "specifies and generalizes the so-called Huygens theorem" — that, in other words, can be interpreted as an expression for the effect of Huygens' wavelets in open regions of an illuminated space. Kirchhoff's friend and now colleague at Berlin, Hermann von Helmholtz, had years before shown how to deploy Green's theorem for the vibrations of sound in open-ended tubes. Moreover, Kirchhoff had already developed the basis of what follows in 1876 in his published lecture on propagation in a compressible fluid.[29]

Consider two scalar functions U and G of x, y, and z *within* a bounded space V whose surface is denoted by s, and whose first and second spatial derivatives are defined and continuous. Let dv be a differential volume element within the space V, ds a differential element of the surface S, and N the normal to ds directed toward the interior of V. Then in Kirchhoff's formulation:[30]

$$\oiint_s ds \left(U \frac{\partial G}{\partial N} - G \frac{\partial U}{\partial N} \right) = \iiint_V dv \left(G \nabla^2 U - U \nabla^2 G \right).$$

Kirchhoff assumed 'initially' that the function G, like φ, satisfied the wave equation. Accordingly, he set U to φ and inserted both it and G into his formulation, replacing the spatial derivatives $\nabla^2 U$ and $\nabla^2 G$ with the time derivatives $\partial^2 U / \partial t^2$ and $\partial^2 G / \partial t^2$ via the wave equation. He then integrated over a time interval, requiring however that the lower limit be negative and the upper limit positive:

$$\int_{-t'}^{t''} dt \oiint_s \left(\varphi \frac{\partial G}{\partial N} - G \frac{\partial \varphi}{\partial N} \right) ds = \frac{1}{a^2} \int_{-t'}^{t''} \left[\iiint_V dv \left(G \frac{\partial \varphi}{\partial t} - \varphi \frac{\partial G}{\partial t} \right) \right] dt.$$
$$(3)$$

Kirchhoff then set G to an expression whose numerator bears a limited resemblance to what would decades later become known as Dirac's delta

---

[28] *Ibid.*, 665.

[29] Kirchhoff 1876, pp. 314–317. There Kirchhoff had used a different specification for the limits in his time integral with attendant changes in the argument which was however less detailed than it later became, perhaps because Kirchhoff wanted to ensure the analysis would work for an infinite train of disturbances (see below).

[30] Kirchhoff 1883, p. 666.

function:

$$G = \frac{F(r_o + at)}{r_o}$$

Here r_o is the distance from any point to an arbitrary but fixed point O, with both points lying within the volume of integration in equation (3). The unusual function F satisfies $F(x) = 0$ for $x \neq 0$, while $\int_d^e F(x) = 1$ where the lower limit d is a *finite* negative number, and the upper limit e is *finite* positive.

This distinguishes Kirchhoff's F from Dirac's delta since the latter's defining integral runs from negative to positive infinity. In his later optical lectures, Kirchhoff justified the existence of such a function as a limiting case by way of an example. Suppose $F(x)$ to be $(\mu/\sqrt{\pi})e^{-\mu^2 x^2}$ and let the constant μ be so large that $F(x)$ is "vanishingly small for every finite value of x" and so as infinite as μ itself when x is zero. Setting $z = \mu x$, then *even integrating over infinity* the function would equal unity because $(1/\sqrt{\pi})\int_{-\infty}^{+\infty} e^{-z^2} dz = 1$, in which case Kirchhoff's F converges asymptotically to Dirac's delta.[31]

Kirchhoff's purpose in introducing the strange 'function' F was to generate surface expressions that could be applied to the physical situation he had in mind, namely the propagation of light within a region that is illuminated by one or more point sources that are *external* to the region. Sections of the bounding surfaces can then be assimilated to light-blocking or reflecting obstacles, allowing as well for different propagatory speeds within different parts of the region in order to accommodate refraction. In doing so, Kirchhoff assimilated the point at which the wave function's value is to be calculated to the point at which the distance r_o in the definition of function G vanishes.

This required two steps. First the volume integral on the right-hand side of equation (3) had to be removed. To do so Kirchhoff imposed a second condition on the lower limit of the time integration, namely that $r_o - at'$ must be "negative and finite". This implies that the distances r_o from O

---

[31] Kirchhoff 1891b, pp. 24–25. Note that the requirement that the integration limits of F must be "finite positive and negative" is maintained. He had first developed an argument for the existence of F in his 1876 derivation for propagation in a compressible fluid. This addendum to the original specification of the function F was added by the editor, Kurt Hensel (1861–1941), of the *Optik* (Kirchhoff 1891b, p. 267), indicating the existence of disquiet concerning the function. We thank Ning Yan Zhu for noting Hensel's intervention.

to any point in the region, thereby including points infinitesimally near the bounding surface, are always less than at'. What might that mean in terms of the physics of the situation? Likely this, though Kirchhoff made no comments: the initial optical disturbance must have begun at a point in time sufficiently distant that it had long passed point O. This presumably licensed Kirchhoff to work thereafter with an effectively continuous train of disturbances of arbitrary form.

From the requirement that $r_o - at'$ must be less than zero (and finite) it followed that the 'function' F must be zero at both limits of the integral, namely $-t'$ and t'', since t'' is positive, and so the entire right-hand side of equation (3) will indeed vanish, leaving only the surface integral on the left. This was not enough for the equation to be made physically applicable. Kirchhoff's unusual 'function' F did wipe out the volume integral, but G still appeared. It did not have any apparent physical meaning, having been chosen for the express purpose of using Green's equation, thereby permitting the consideration of surfaces and hence the general behavior of light in the presence of physical boundaries. Moreover, the inclusion of point O involves a singularity since r_o vanishes there. This is why Kirchhoff decided to cut O altogether out of the volume over whose surfaces the integration takes place.

To that end Kirchhoff surrounded O with an "infinitesimal sphere" which he subtracted from his region. That produced two bounding surfaces: the original one as external boundary, and the spherical surface surrounding O as internal boundary (Figure 7). The result was to split the surface integral into two parts, one over the external, the other over the internal boundary. With lowercase s denoting the external, and uppercase S the internal boundary, Kirchhoff's equation becomes:[32]

$$\int_{-t'}^{t''} dt \oiint_{s+S} \left(\varphi \frac{\partial G}{\partial N} - G \frac{\partial \varphi}{\partial N} \right) d(s+S)$$

$$= \int_{-t'}^{t''} dt \oiint_{s} \left(\varphi \frac{\partial G}{\partial N} - G \frac{\partial \varphi}{\partial N} \right) ds + \int_{-t'}^{t''} dt \oiint_{S} \left(\varphi \frac{\partial G}{\partial N} - G \frac{\partial \varphi}{\partial N} \right) dS = 0$$

$$(4)$$

---

[32] *Ibid.*, pp. 666–668.

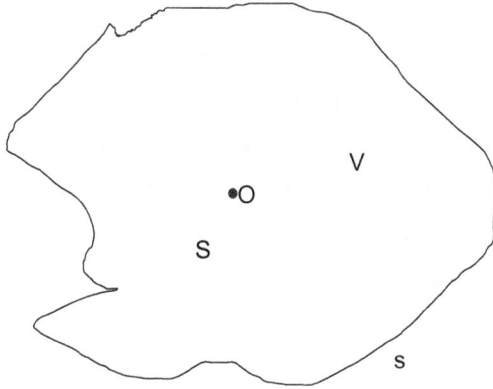

Figure 7:   Kirchhoff's configuration for Green's theorem.

Kirchhoff next moved to eliminate G. First of all, r_o now becomes the radius R of the "infinitesimal sphere," while its surface element dS has the factor R^2. Consequently Gds is proportional to R which, since the sphere is infinitesimal by assumption, means that the second term in the integral over the sphere's surface S can be discarded. Taking however the normal gradient of G at the sphere's surface yields $-(1/R^2)F(R+at)$, which is $-(1/R^2)F(at)$ since R can be neglected in the function's argument. Multiplication by dS accordingly leaves just $F(at)$, and so the integral over S simply reduces to $-4\pi\varphi_0$ where φ_0 is the value of the wave function at the locus of the fixed point O at time zero.[33] To eliminate F altogether, Kirchhoff recurred to his requirement that its integral over any finite interval must be equal to one, in which case integrating $F(at)$ over time simply produces $1/a$, with equation (4) thereby reduced to the following:

$$\int_{-t'}^{t''} dt \oiint_{s} \left(\varphi \frac{\partial G}{\partial N} - G \frac{\partial \varphi}{\partial N} \right) ds = \frac{-4\pi}{a} \varphi_0(0) \tag{5}$$

Note that the locus of whatever luminous point gives rise to the source wave is not as yet under direct consideration in Kirchhoff's analysis beyond requiring that it must lie outside the bounded region in order to avoid a further singularity.

Both G and a time integral still appear here and had to be removed in order to reach a physically useful result — in order, that is, to calculate

---

[33] At time zero because $F(at)$ is itself non-zero only when its argument vanishes.

the amplitude at a given point in the region bounded by the surface of integration. The region in question must not at this stage contain a source. Again using the integral requirement over the finite interval $[t', t'']$ and now shifting the time origin to t, Kirchhoff reduced equation (5) to the following form, which we shall call his fundamental result:[34]

$$4\pi\varphi_0(t) = \oiint_s ds \cdot \Omega$$

where

$$\Omega = \frac{\partial}{\partial N} \frac{\varphi(\vec{r}_o, t - r_o/a)}{r_o} - \frac{f(\vec{r}_o, t - r_o/a)}{r_o} \quad \text{and} \quad f = \partial\varphi/\partial N. \quad (6)$$

Helmholtz, as Kirchhoff knew, had derived a similar formula in 1859 to characterize the acoustic vibration within a pipe with one open end, albeit in a considerably less general manner. Unlike Kirchhoff, Helmholtz assumed temporal harmonic variation, thereby separating the time- from the space-dependent part of the wave function. Kirchhoff, recall, had not required such a limitation, which is why he had introduced his function F.[35] Note that in Kirchhoff's formulation the loci of whatever luminous points are responsible for the radiation do not appear explicitly. The waves that they engender will appear only by supposition in the boundary conditions that Kirchhoff adopted. Kirchhoff did not, in other words, separately develop Green integrals for the incident and resultant waves and then relate them through his boundary conditions.

Kirchhoff now interpreted his fundamental result explicitly in terms of Huygens' principle: "the motion of the ether [at any point] in the space enclosed in surface s can be regarded as caused by a layer of luminous points on surface s, because each one of the two terms of which Ω is composed may be described as corresponding to a luminous point situated at

---

[34] *Ibid.*, 668–669.

[35] H. v. Helmholtz: Theorie der Luftschwingungen in Röhren mit offenen Enden, *Journal für die reine und angewandte Mathematik*, 62:1 (1859), 23. Suppose with Helmholtz that $\varphi(\vec{r}, t) = \varphi_c(\vec{r}) \cos(\omega t) + \varphi_s(\vec{r}) \sin(\omega t)$ so that the time and space dependencies could be separated. Helmholtz's equation may then be written as follows:

$$4\pi\varphi_{c,s}(r_o = 0) = \oiint_S ds \cdot \left\{ \frac{\partial\varphi_{c,s}}{\partial N} G_{c,s}(r_o) - \varphi_{c,s} \frac{\partial G_{c,s}(r_o)}{\partial N} \right\}$$

with

$$G_c(r) = \frac{\cos(kr)}{r}, \quad G_s(r_o) = \frac{\sin(kr)}{r}.$$

the location of *ds*."[36] However, where Huygens had based his principle on a consideration of the physical tendency to expand at each point of a wave front, Kirchhoff's version emerged as a result of his application of Green's theorem without any direct consideration of the physics beyond the wave equation proper. Of primary significance, his expression equation (6) indicated that the contribution from each point on an arbitrary surface *s* to the ether vibration at a point *O* that is enclosed by *s* is determined by both the ether vibration φ and its normal gradient on *s*. Both values had consequently to be specified.

Kirchhoff recognized that his result could be extended to a case in which the point O is *exterior to* the bounding surface *s* while the originating luminous points (which are thus far exterior to *s*) lie *within* the enclosed region, amounting thereby to an inversion of the existing situation. Key to the first step in the derivation of this extension was to redefine the space exterior to *s* as a volume bounded internally by *s* and externally by the infinitely large surface S_∞ so that (6) becomes:

$$4\pi\varphi_o(t) = \oiint_s ds \cdot \Omega + \oiint_{S_\infty} dS_\infty \cdot \Omega.$$

Here the normal to *s* is turned from inward to outward. Supposing a "particular condition" to be satisfied — namely, that before some finite past moment the values of both functions (φ and f) were everywhere zero — then, provided that the point O is not itself at infinity and choosing a subsequent but otherwise arbitrary moment, both functions will at that time also vanish on the infinitely large surface S_∞.[37] This licensed a principle of inversion, according to which whatever holds for a point O applies as well to a luminous point by exchanging their locations. In such an exchange the locus of evaluation for the wave remains near the Green function singularity.

To further extend the situation, Kirchhoff envisioned two closed surfaces that enclose volumes which have an intersection within which point O lies, while any luminous points *I* lie outside both regions (Figure 8). Each of these surfaces then yields ϕ_0 near O. Subtracting the intersection from the

---

[36] *Ibid.*, pp. 669–670. The translation is by Ann Hentschel (2016) (in this volume, pp. 75–76).

[37] Kirchhoff's requirement was in later years subsumed under what became known as the Sommerfeld radiation condition, according to which the limit at infinity of the difference $\partial\phi/\partial N - 2\pi i\phi/\lambda$ must vanish: *cf* Goodman 1988, p. 44.

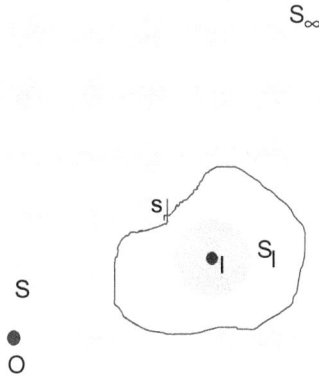

Figure 8: Kirchhoff's further configuration for Green's theorem.

union of the volumes creates a region with a new surface that bounds a region from which both O and any luminous points are excluded (because O now lies within the extracted intersection of the two original volumes). Under theses circumstances the integral in equation (6) always vanishes. By the principle of inversion the same follows if the original two surfaces enclosed luminous points, with O lying outside both. A similar procedure, but this time forming the composite union of two intersecting surfaces, implies that the integral in equation (6) also vanishes if a surface *includes* both O and I. In summary Kirchhoff has demonstrated:

1. a region that *includes* O but *excludes* I can produce a non-zero value for the integral in equation (6). The same holds *vice versa* by inversion.
2. a region that *excludes* both O and I produces a zero value for the integral in equation (6).
3. a region that *includes* both O and I also produces a zero value for the integral in equation (6).

These three correspond respectively to cases in which (1) I and O are separated by a surface, (2) the radiation from I simply passes through the region, and (3) I's radiation reaches O unimpeded.

Before any specific results could be obtained, Kirchhoff had to produce a useful expression for the general integral, $\iint_s ds \cdot \Omega$, given the locus of

O and the waves engendered by whatever luminous points are present. To do so, he assumed a single radiating point I and limited the form of the wave that it produces to a time harmonic expression, namely to $\varphi = (1/r_i)\cos[(r_i/\lambda - t/T)2\pi]$. The distances r_o, r_i that now both appear in the general integral are respectively drawn from the loci O, I to points on the surface of integration. This produced the following form for the integrand Ω:

$$\Omega = \frac{1}{r_i r_o}\left(\frac{1}{r_i}\frac{\partial r_i}{\partial N} - \frac{1}{r_o}\frac{\partial r_o}{\partial N}\right)\cos\left[\left(\frac{r_i + r_o}{\lambda} - \frac{t}{T}\right)2\pi\right]$$
$$+ \frac{2\pi}{r_i r_o \lambda}\left(\frac{\partial r_i}{\partial N} - \frac{\partial r_o}{\partial N}\right)\sin\left[\left(\frac{r_i + r_o}{\lambda} - \frac{t}{T}\right)2\pi\right].$$

Kirchhoff next introduced a specific coordinate system and an approximation linked to it. The origin of the new coordinate system was placed on the surface of integration at a very specific point: namely, where the sum $r_o + r_i$ is a minimum and therefore is stationary. This will occur if the line joining I to O passes through the surface of integration, in which case the origin will lie at the intersection of the joining line with the integration surface. Kirchhoff's new origin is effectively the same as the one that Fresnel had introduced decades before, namely the pole. Using this coordinate system, he expanded r_o, r_i as functions of ρ_0, ρ_1, these being the distances from O, I to the pole, but only up to second order. That limitation amounted to the requirement that the wavelength is vanishingly small in comparison to the distances r_o, r_i.[38] This would in contemporary parlance be referred to as the "stationary phase approximation."

For a completely unobstructed wave Kirchhoff could then show that integrating the resultant expression simply reproduces the presumptive wave from the luminous point because the normal gradients that appear in the sine term simply cancel one another, obliterating it, leaving only the cosine term. He was now ready to grapple with the general problem of

---

[38] Kirchhoff expressed this as the wavelength being 'infinitesimal' and the sum $r_o + r_i$ being effectively constant throughout the integration since the limitation to a second-order expansion restricts the integration's accuracy to loci within the vicinity of the origin, where the sum is a minimum (Kirchhoff 1883, p. 672).

optics, including retrieving the situation that would reproduce a facsimile of an optics based on bundles of rays. To do so required a crucial physical assumption, known later as "Kirchhoff's *Ansatz*" or as his "physical optics approximation": namely, that the light wave ϕ_i (and its gradient $\partial \varphi_i / \partial N$) at any point on the surface of integration is approximately equal to the source wave from I (and its gradient) plus any reflected wave ϕ_r (and its gradient $\partial \varphi_r / \partial N$), meaning that the presence of any obstructing body does not alter the wave that is incident upon it other than by affecting the values of the wave function and its normal gradient over the body's actual surface:[39]

$$\varphi \cong \varphi_i + \varphi_r, \quad \frac{\partial \varphi}{\partial N} \cong \frac{\partial \varphi_i}{\partial N} + \frac{\partial \varphi_r}{\partial N}.$$

Kirchhoff considered three specific situations: light scattering from a "black body", light diffracted from a black-body screen with an aperture, and diffraction by a grating. For our purposes here, we will consider only his treatment of diffraction. A light source at point i is enclosed by an opaque screen whose surface S is punctured by an aperture A. To simplify the calculation, Kirchhoff transformed the screen from an enclosed surface to an infinite plane, so that the light source i and the point of observation O were on opposite sides of the infinite, flat screen. Here we encounter Kirchhoff's requirements for *both* the wave function at the screen's two surfaces and its normal gradients there. Divide S into a part S_i that faces the light source and a part S_o that is shielded from it. Since the screen extinguishes all light striking it, the wave function must vanish altogether over S_o. In addition, the wave over the aperture A should have the same form as the source wave φ_i according to Kirchhoff's *Ansatz*, which leaves the wave near S_i (and so over the aperture) unaltered if the screen is black (there being no reflection). Note Kirchhoff's explicit consideration of both sides of the screen.

However, neither the *Ansatz* nor Kirchhoff's definition of an opaque screen in themselves placed any limitation on the wave's normal gradient over S_o. Kirchhoff nevertheless assumed that it too would vanish over the shadowed surface of the screen, producing the following set of boundary

---

[39] *Ibid.*, pp. 683–685.

conditions:[40]

$$\varphi = \varphi_i, \quad \frac{\partial \varphi}{\partial N} = \frac{\partial \varphi_i}{\partial N} \quad \text{on } A$$

$$\varphi = 0, \quad \frac{\partial \varphi}{\partial N} = 0 \quad \text{on } S_o.$$

Using these boundary conditions with a source wave set to $(1/r_i)\cos[(r_i/\lambda - t/T)2\pi]$, the integrand Ω produces the following result for the wave at an observation point O under the approximation that the wavelength is vanishingly small in comparison with the distances of both the source and observation points from the aperture:[41]

$$\varphi_o = \iint_A ds\Omega = \frac{1}{2\lambda} \iint_A \frac{ds}{r_i r_o} \left(\frac{\partial r_i}{\partial N} - \frac{\partial r_o}{\partial N} \right) \sin \left[2\pi \left(\frac{r_i + r_o}{\lambda} - \frac{t}{T} \right) \right].$$

The term containing the difference between the normal gradients of $r_{o,}$, r_i required two further approximations: namely that these distances are much larger than the dimensions of the aperture and so can be considered constant across it outside of the sine term in the integrand, and that the line from I to a point in the aperture forms an "infinitesimal angle" with the line from O to that point. This latter limitation in effect means that the observation point O must lie within or near the cone that has I at its apex and the aperture as its boundary, i.e. that O lies within or near what in the absence of diffraction would be the geometric shadow of

---

[40] Given these boundary conditions, and a special condition, Kirchhoff could also retrieve geometric optics. Require that the sum of the distances from I and O to the surface must be constant for any finite part of it — meaning in effect that both the observation and luminous points are very far away. If, under this condition, the line joining I to O does not anywhere intersect a body that does not reflect light — a "black body" — then the wave at O remains unaltered. If, on the other hand, that line passes through the black body at least once, then "darkness occurs at the location of O", a true shadow is formed and in consequence "the light from the luminous point propagates rectilinearly in rays that can be regarded as independent of one another" (*ibid.*, pp. 686–687). This retrieves geometric optics for a black-body obstacle and an effectively infinitesimal wavelength. The limitation to constancy over a finite area of the surface for the sum $r_o + r_i$ is dropped for diffraction.

[41] *Ibid.*, p. 688. Note that $\partial r_i/\partial N = \cos(\vec{r}_i, \vec{n}_i)$ and $-\partial r_o/\partial N = \cos(\vec{r}_o, \vec{n}_o)$, where $\vec{r}_i$ is the vector from an arbitrary point on A to the point of illumination I, $\vec{r}_o$ is the vector from the same point on A to the point of observation O, $\vec{n}_i$ is the vector normal to A and pointing toward the side of I, $\vec{n}_o$ is the vector normal to A and pointing toward the side of O, and $\cos(\vec{A}, \vec{B})$ is the cosine of the angle between the two vectors $\vec{A}$ and $\vec{B}$. Kirchhoff's diffraction integral may therefore be written as $\varphi_o = \frac{1}{2\lambda} \iint_A \frac{ds}{r_i r_o} [\cos(\vec{r}_i, \vec{n}_i) + \cos(\vec{r}_o, \vec{n}_o)] \sin \left[2\pi \left(\frac{r_i + r_o}{\lambda} - \frac{t}{T} \right) \right]$.

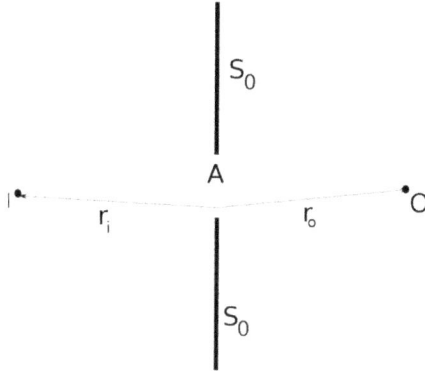

Figure 9:   Kirchhoff's diffraction configuration.

the luminous point. In that case the normal gradients in Ω become equal and opposite since they are simply the cosines of the angles that the lines from I and O make with the normal, while the product r_o, r_i is effectively constant, yielding the following expression for the wave at O:

$$\varphi_o = \frac{1}{\lambda r_i r_o} \frac{\partial r_i}{\partial N} \iint_A ds \sin\left[2\pi\left(\frac{r_i + r_o}{\lambda} - \frac{t}{T}\right)\right]$$

Note immediately one difference from Fresnel's original expression: the Huygens wavelets are shifted by a quarter-wavelength in phase from the source wave, which generated discussion in subsequent years.[42]

Kirchhoff added an incident time-harmonic sine wave with a different amplitude for generality and could at once write down an expression for the optical intensity at O. For simplicity we will assume a unit total incident intensity in which case Kirchhoff's diffraction integral yields:[43]

$$\text{intensity at } O = \frac{1}{2\lambda^2 r_i^2 r_o^2}\left(\frac{\partial r_i}{\partial N}\right)^2 (c^2 + s^2)$$

---

[42]Fresnel took the wave in Huygens' principle to differ from the source wave solely by virtue of distance to the surface of integration, so that, e.g., a cosine wave remained a cosine wave plus a phase addition to its argument. However the integrands in Kirchhoff's expression are shifted by a quarter wavelength from the source wave in addition to the phase addition. The shift is a direct consequence of applying Green's theorem to the wave equation.

[43]Kirchhoff 1883, p. 689.

where

$$c = \int ds \cos\left(\frac{r_i + r_o}{\lambda}\right) 2\pi \quad \text{and} \quad s = \int ds \sin\left(\frac{r_i + r_o}{\lambda}\right) 2\pi.$$

The functions c, s are of course the Fresnel integrals, while the factor in the square of the normal gradient of r_i is Kirchhoff's inclination factor, here obtained for the first time directly from the solution to the wave equation *via* Green's theorem and suitable approximations. Except for this factor multiplying the integrals, the result has the same form as the one that Fresnel had produced decades before for the same situation. Kirchhoff's is more general in two ways: first it includes a factor that depends directly on an inclination factor and inversely as the product of the wavelength by the distances of I, O to the aperture loci, and second Kirchhoff left his result in terms of r_o, r_i instead of further approximating in terms of the vertical distances of I, O to the screen.

5. The Poincaré Paradox

Trained in engineering at the *École Polytechnique* and *École des Mines*, Poincaré had always kept an active line of research in mathematical physics and applied mathematics, in spite of his better known works in more abstract areas. In 1886 he assumed the chair of mathematical physics at the University of Paris, where he lectured on optics in the winter semester of 1887–1888, which he again taught in 1891–1892.[44] Both sets were published, and in the 1892 lectures Poincaré reviewed what he then termed "Kirchhoff's hypotheses" for diffraction.[45] If, he began, two functions G, φ are continuous and finite within a given region, then Green's theorem requires:

$$\iint_{\Sigma} \left(\varphi \frac{\partial G(r)}{\partial N} - G(r)\frac{\partial \varphi}{\partial N}\right) d\sigma = \iiint_{V} dv(G(r)\nabla^2\varphi - \varphi\nabla^2 G(r)),$$

where the surface of integration is denoted by Σ. Suppose next that G has a singularity at the origin O of coordinates but such that the product Gr

---

[44] For a thorough account of Poincaré's life and career see Gray 2013. Also Dieudonné 2008, Charpentier, Ghys *et al.* 2010. On anomalous dispersion and its theoretical consequences see Buchwald 1985, chap. 27.
[45] Poincaré had derived formulae akin to Kirchhoff's in the 1887–1888 lectures, though at that time he was unaware of Kirchhoff's theory; he had then also noted that both boundary conditions cannot strictly hold simultaneously: Poincaré 1889, pp. 115–116. On Poincaré and light see Darrigol, 2015.

becomes unity there, with r denoting distance from O. Surround O with an infinitesimal sphere and calculate the integral over its surface, producing (as with Kirchhoff) the value $-4\pi\varphi_O$, which gives the value for φ at the locus O of the singularity for G. Assuming further that both φ and G satisfy the wave equation, then the volume integral vanishes, leaving Poincaré with equation (7) for the value of φ_O at a point that lies within an infinitesimal sphere which is itself surrounded by a closed surface Σ:[46] If point O is not surrounded by Σ, then the integral vanishes.

$$\oiint_{\Sigma} \left(\varphi \frac{\partial G(r_P)}{\partial N} - G(r_P) \frac{\partial \varphi}{\partial N} \right) d\sigma = -4\pi\varphi_O. \tag{7}$$

To apply the theorem to diffraction, Poincaré considered a situation that was similar to Kirchhoff's but that was configured in a different manner. He divided space into four distinct surfaces as follows. Marking an arbitrary point that we shall designate as M, Poincaré denoted the surface of the screen facing M as B and the screen's opposite surface as C. He then described a closed surface S, part of which coincides with B and within which point M lies; A denotes the part of S that excludes B. Surface S is consequently the union $A + B$. Finally, he surrounded M with an infinitesimal sphere s_M that lies within the region surrounded by S. We have added a point P that lies entirely outside the union $C + A$ and have surrounded it, like M, with an infinitesimal sphere s_p. Poincaré's configuration placed the locus of his luminous point at M within the region surrounded by the surface, which is where Kirchhoff had placed his observation point. The observation point, though Poincaré did not mark it, here lies at P, outside $C + A$.

Poincaré introduced a function ϕ^i to represent the wave at any point emitted by the luminous point M in the absence of a physical screen. When B and C are the surfaces of a physical object, then the wave φ^i is altered by their presence to φ^R according to Kirchhoff's boundary conditions. Poincaré employed the distinction to establish relations between the two functions using the boundary conditions for an opaque body. Kirchhoff had not done anything similar since he had in the end directly inserted the value of the wave from the luminous point into his integral and had applied boundary conditions without having reached the result by means of relations between distinct Green's integrals for φ^i and φ^R.

---

[46] Poincaré 1892b, pp. 141–143.

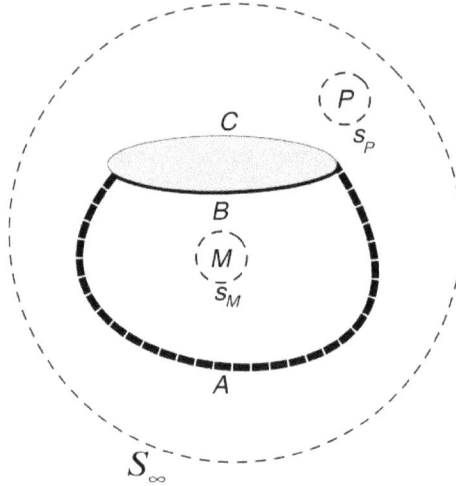

Figure 10: Poincaré's first surfaces.

To clarify Poincaré's somewhat terse discussion we introduce the following convention: the value φ_P at a point P of the integral $\frac{1}{4\pi} \iint_{\sigma_k} \left[G_P \frac{\partial \varphi}{\partial N} - \varphi \frac{\partial G_P}{\partial N} \right] d\sigma$ of a function φ and a Green function G_P, the latter of which has a singularity at P, over a set of surfaces σ_k that bound the region enclosing P will be represented by $\iint_{\sigma_k} [\varphi]$.[47] In order to employ Green's theorem for the value of the wave function at P we surround P with an infinitesimal sphere in Kirchhoff's fashion so that we can use Green's theorem to replace the integral over the surface of that sphere with the value of the wave function at P. Then, with reference to Figure 10, we can with Poincaré build expressions for both φ^i and φ^R by considering different sets of the surfaces. Consider first a surface formed by the union of s_M with the surfaces B, C. A source wave φ_P^i from M that reaches the point P within the region bounded by these surfaces will accordingly be represented by the following expression:

$$\varphi_P^i = \iint_{s_M + B + C} [\varphi^i]. \tag{8}$$

---

[47]Poincaré always presumed an outer surface S_∞ at infinity at which the wave function φ and its normal derivative vanish, with the surfaces σ_k forming inner boundaries. The Green's function at point U is $G_p = \exp\left(\frac{-ikr_p}{r_p} \right)$, where r_p is the distance between U and P, $k = \frac{2\pi}{\lambda}$ and the time dependence has been removed due to the monochromatic assumption.

The resultant wave φ_P^R at P can be similarly represented by integrals over these same three surfaces:

$$\varphi_P^R = \iint_{s_M+B+C} [\varphi^R]. \tag{9}$$

However, the latter is affected by Kirchhoff's boundary conditions, for which Poincaré assumed the region surrounded by B and C to be opaque. Accordingly, and assuming Kirchhoff's *Ansatz*, the source wave at surface B, which faces the luminous point at M, is unaffected by the presence of the body, while at surface C Kirchhoff's boundary conditions required the resultant wave and its normal derivatives to vanish. Consequently φ_P^R can be expressed in terms of the source wave and just the two surfaces B, s_M:

$$\varphi_P^R = \iint_{s_M+B} [\varphi^i] = \iint_{s_M+B+C} [\varphi^i] - \iint_C [\varphi^i]. \tag{10}$$

We next follow Poincaré by introducing a new set of boundaries using the surfaces in expression (10), specifically the space bounded internally by the union of A with C (and externally by the surface at infinity). The source wave at any point in this region consequently has the expression:

$$\varphi_P^i = \iint_{A+C} [\varphi^i]. \tag{11}$$

Equating the expressions (8) and (11) for φ_P^i produces expression (12):

$$\varphi_P^i = \iint_{s_M+B+C} [\varphi^i] = \iint_{A+C} [\varphi^i]. \tag{12}$$

As a result, expression (10) for the diffracted wave φ_P^R becomes expression (13):

$$\varphi_P^R = \iint_{s_M+B+C} [\varphi^i] - \iint_C [\varphi^i] = \iint_{A+C} [\varphi^i] - \iint_C [\varphi^i] = \iint_A [\varphi^i]. \tag{13}$$

That is, the diffracted wave at a point P can be found by integrating the wave from the luminous point M over an open surface whose unclosed portion is bridged by the obstacle, with M lying within the region enclosed by the bridged surface and P lying outside it, precisely as Kirchhoff had found.

For clarity, we can redraw as follows to represent the typical case of a screen with an aperture. If we imagine the obstacle bounded by B and C to be an infinite plane screen and the open surface A to be an aperture on the screen, then Poincaré's configuration in Figure 11 is nothing but the

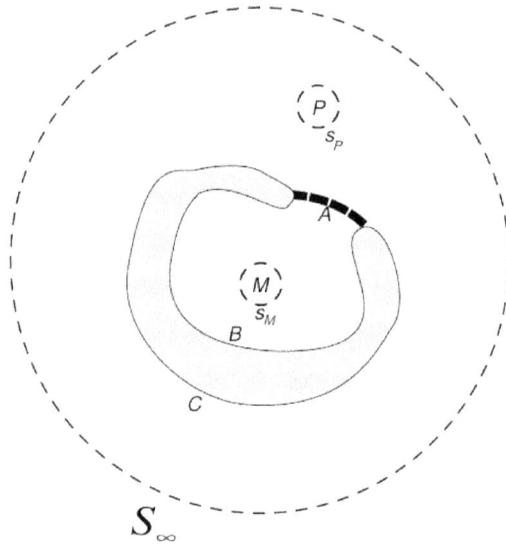

Figure 11: Poincaré's surfaces adapted to an opaque screen $(B + C)$, with an aperture (A), enclosing a luminous point (M) and an observation point (P).

configuration for Kirchhoff's diffraction problem (surface C in that case becomes the screen surface facing away from Poincaré's luminous point M, i.e. C becomes the shadowed surface).

Poincaré's explicit introduction of integrals for the source wave revealed a problem that Kirchhoff would not have seen precisely because he had considered the source wave only towards the end of his calculations — until then his luminous point, though necessarily present, had not directly entered. With Poincaré we now consider a point L located *within* the region $B + C$ bounded by the surfaces of the screen itself. This point is the locus of a Green's function G_P. In doing so we first consider B and C merely as surfaces and not as the boundaries of an opaque screen in order to produce an expression for the source at L that will lead us to the difficulty that Poincaré discovered.[48]

Recall first that in these calculations a surface S_∞ at infinity is presumed where the wave function and its normal derivatives vanish. We now develop expressions for the values of the source wave and of the total wave at L by

---

[48] Poincaré's deduction of the difficulty is extremely terse. What follows draws out the several relations that he developed in somewhat greater detail. Also see Baker & Copson 1939, pp. 70–72.

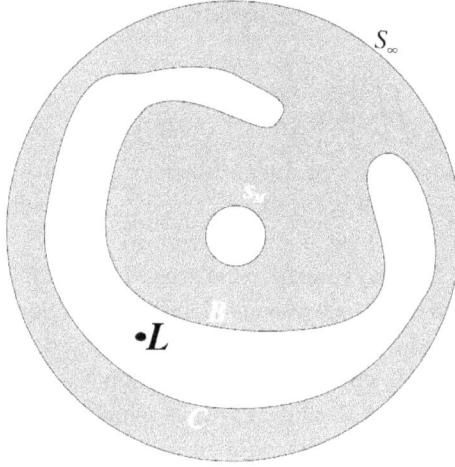

Figure 12: Poincaré's point L within the region bounded by $B + C$.

integrating *within the region that does not enclose it* — the darkened area in Figure 12, which is bounded by s_M, S_∞, and the surfaces B, C of the screen itself.

Since point L does not lie within the region of integration, and since there are no singularities for the source wave, the total wave, or the Green's function G_L within that region, then Green's theorem requires that the integral for either the source wave or for the total wave over the boundaries and using the Green's function G_L must vanish (since the values also vanish at infinity by assumption we may as always neglect that boundary):

$$\iint_{s_M+B+C} \left[G_L \frac{\partial \varphi^i}{\partial N} - \varphi^i \frac{\partial G_L}{\partial N} \right] d\sigma = 0, \tag{14}$$

$$\iint_{s_M+B+C} \left[G_L \frac{\partial \varphi^R}{\partial N} - \varphi^R \frac{\partial G_L}{\partial N} \right] d\sigma = 0. \tag{15}$$

However — the key element in Poincaré's proof — Kirchhoff's boundary conditions reduce equation (15) to equation (16) *since φ^R and its normal derivatives should both vanish over C*, while taking the values of φ^i over B and s_M.

$$\iint_{s_M+B} \left[G_L \frac{\partial \varphi^i}{\partial N} - \varphi^i \frac{\partial G_L}{\partial N} \right] d\sigma = 0. \tag{16}$$

Subtracting equation (16) from equation (14) accordingly yields equation (17):

$$\iint_C \left[G_L \frac{\partial \varphi^i}{\partial N} - \varphi^i \frac{\partial G_L}{\partial N} \right] d\sigma = 0. \tag{17}$$

And here we find Poincaré's problem. Since φ^i is an arbitrary source wave, and since surface C is itself not only arbitrary but also without physical constraints when considering the source wave, equation (17) would imply that such a wave would have to vanish "whatever the form of the screen". Waves could simply not exist in the presence of a diffractor. Something was clearly wrong with the boundary condition over the shadowed surface of the screen, since it is that condition which produces equation (16). Poincaré did not also deduce through an example that the wave which results from Kirchhoff's boundary conditions cannot reproduce the very conditions over the aperture proper that led to it (*cf* footnote 49 below). He had instead found that waves could simply not occur in diffraction given Kirchhoff's conditions.

In working with Green's theorem earlier in his lectures, Poincaré had remarked that "in general, it will not be possible arbitrarily to assign to one a system of values, whether of φ or $d\varphi/dN$, because these two functions are linked by a relation expressing that the integral is null in an exterior point."[49] This is perhaps what led him to investigate what occurs when the integrations are carried out for "an exterior point" in Green's theorem, leading to the contradiction between the existence of waves and Kirchhoff's boundary conditions that he uncovered. Poincaré's remark concerning overdetermination would certainly have been clear to most mathematicians and physicists in his day. However, that fact alone does not lead to his result, which requires considerably more: the problem arises because the total wave (and its normal derivative) is presumed to be the same as the source wave over the entire visible part of the obstacle (surface B in Figure 12 Poincaré's point L within the region bounded by $B + C$), for that was why φ^i appeared in Poincaré's several expressions for φ^T, but that it must vanish on the shadowed surface (C).

In 1897, Lord Rayleigh in England took an entirely different tack. He did not mention the Poincaré paradox at all, but instead developed two alternative expressions for the diffracted wave. Neither of the two solutions suffered in the manner that Poincaré had pointed out because, unlike Kirchhoff's expression, Rayleigh's alternatives did not impose simultaneous

---

[49] Poincaré 1889, pp. 144–145.

requirements on both the wave and its normal derivative — which Poincaré had himself pointed to as a possible source of the problem, though we have seen that it required both that *and* a discontinuity of the wave and its normal derivative in crossing from the unshielded to the shielded region.

6. The Rayleigh–Sommerfeld Alternatives

By the mid-twentieth century, the inconsistency of Kirchhoff's theory seemed to many to be an obvious consequence of the well-known boundary conditions for partial differential equations. As E.W. Marchand (1914–1999) at Eastman Kodak's Research Laboratory and Emil Wolf (1922–) at the University of Rochester (the latter of whom had studied under Max Born) remarked in 1966:

> Since the time-independent wave equation for U is elliptic, the specification of U or its normal derivative $\partial U/\partial n$ in the plane of the aperture (together with the specification of the asymptotic behavior of the solution at infinity in the appropriate half-space) is sufficient to specify the solution uniquely. The inconsistency in Kirchhoff's theory arises from the specification of both U and $\partial U/\partial n$ and has the consequence that, as the point of observation P approaches the plane of the aperture, the Kirchhoff solution $U_K(P)$ does not recover the assumed boundary conditions.[50]

It had for decades been common knowledge that solutions to hyperbolic or elliptic partial differential equations are fully determined by either, but not both, of the following two conditions: (i) the Dirichlet, which specifies the value of the function on a given boundary, or (ii) the Neumann, which specifies the value of the function's normal derivative there.[51] From this perspective, of which we have just seen that Poincaré was quite aware, Kirchhoff's theory must be problematic. Yet the result seemed to work quite well empirically given the approximations that Kirchhoff had made. Poincaré accordingly thought that, though the two requirements are not "rigorously compatible, they are at least so in an approximate manner,

---

[50] Marchand & Wolf 1966, p. 1712.

[51] Hyperbolic equations, of which the time-dependent wave equation is one, have the form $\partial^2 \varphi/\partial x^2 - (1/c^2)\partial^2 \varphi/\partial t^2 = f(x,t)$ while elliptic equations satisfy $\partial^2 \varphi/\partial x^2 + \partial^2 \varphi/\partial y^2 = f(x,y)$ — Laplace's, Poisson's and the time-independent wave equation developed by Helmholtz are all elliptic.

when one neglects quantities of the order of a wavelength."[52] He did not however provide any clear justification for such a claim.

In 1897 Rayleigh, the leading expert on the theory of sound (having published the first edition of his magisterial treatise on the subject in 1877) detailed a theory that avoided overspecifying the boundary conditions.[53] Rayleigh examined two situations: in one (φ^I) the normal gradient of the wave vanishes over the shadowed surface of the screen, while in the other (φ^{II}) the wave itself does. In both cases the wave and its normal gradient must be continuous across the aperture. This produces a duo of possible solutions as follows, wherein S denotes the surface of the screen including any aperture and A denotes an aperture:[54]

$$\varphi^I = \frac{-1}{4\pi} \left[\iint_{S-A} ds \left\{ \varphi \frac{\partial G}{\partial N} \right\} + \iint_A ds \left\{ \varphi \frac{\partial G}{\partial N} - G \frac{\partial \phi}{\partial N} \right\} \right],$$

$$\varphi^{II} = \frac{1}{4\pi} \iint_{S-A} \left[ds \left\{ G \frac{\partial \varphi}{\partial N} \right\} + \iint_A ds \left\{ \varphi \frac{\partial G}{\partial N} - G \frac{\partial \phi}{\partial N} \right\} \right].$$

This division into two situations raised the problem of how to find distinct expressions for φ in each case. The issue did not arise with Kirchhoff's integral across the aperture because he dropped the integral across the screen proper. Rayleigh solved the problem by considering two cases. For φ^I he envisioned a *perfectly reflecting* plane screen on which a plane wave parallel to it is incident. He then offered expressions χ_m, χ_p respectively for what such a wave would be on either side of the screen when it is *unperforated*, so that for a screen S in the yz plane $\chi_m = e^{-ikx} + e^{ikx}$ and $\chi_p = 0$. To these he added, again respectively, $\psi_m = \iint_A \Psi_m \frac{e^{-ikr}}{r} dS$, $\psi_p = \iint_A \Psi_p \frac{e^{-ikr}}{r} dS$, integrating solely over an aperture A, that are presumed to modify these solutions to take account of the perforation. The functions Ψ_m, Ψ_p must satisfy the reduced wave equation.

For continuity conditions in φ^I, again, Rayleigh needed the normal gradient of the wave over the shadowed surface of the screen to vanish, but for the wave as well as its normal gradient to be continuous across the aperture,

---

[52] Poincaré 1892b, p. 188.
[53] Rayleigh 1897. The details of Rayleigh's analysis depended upon results he had developed in his *Theory of Sound*, whose second edition of the first volume had appeared in 1894 and, for the relevant parts of the second volume, the year before.
[54] Rayleigh 1896, secs. 277, 278 and 292, the last of which Rayleigh referred to in his paper on optical diffraction, Rayleigh 1897.

i.e. at the screen, where $x = 0$, Rayleigh required in φ^I that:

$$\text{on } S - A: \quad \frac{\partial(\chi_p + \psi_p)}{\partial N} = 0$$

$$\text{on } A: \quad 2 + \psi_m = \psi_p \quad \text{and} \quad \frac{\partial(\chi_m + \psi_m)}{\partial N} = \frac{\partial(\chi_p + \psi_p)}{\partial N}.$$

Since χ_p (and so its normal gradient) is required to vanish over the entire surface S, and the normal gradient of χ_m also vanishes over S (where $x = 0$), these requirements can be satisfied provided that Ψ_m, Ψ_p are equal and opposite on A, with the wave functions ψ_m, ψ_p then being equal and opposite by reflection of the one across the screen to produce the other, with the additional stipulation that the wave function ψ_m must have the value -1 on the aperture proper. The problem then consists, as Rayleigh put it, "in so determining Ψ_m that this shall be the case." Conversely, in the case of φ^{II}, where the wave is presumed to vanish over the shadowed surface S-A, Rayleigh altered χ_m to the difference $e^{-kx} - e^{kx}$ instead of the sum, so that both χ_m and χ_p vanish everywhere at the screen. In this situation ψ_m, ψ_p are, respectively, $\iint_A \Psi_m \frac{\partial(e^{-ikr}/r)}{\partial N} dS$, $\iint_A \Psi_p \frac{\partial(e^{-ikr}/r)}{\partial N} dS$. To satisfy continuity over the aperture then required Ψ_m, Ψ_p to be equal and opposite there, where Ψ_m takes a value such that $\partial\psi_m/\partial N$ becomes *ik*.

Note, then, that Rayleigh's duo each in its specific manner only partially satisfy Kirchhoff's boundary conditions over the body S-A of the screen: in the case of φ^I the wave function over S-A must vanish, whereas its normal gradient need not, whereas in the case of ϕ^{II} the reverse holds. The problems then require judicious choices of the wave functions and appropriate conditions at the aperture using the same Green's function $\frac{e^{-ikr}}{r}$. Rayleigh's aim had not been to avoid the inconsistency that plagued Kirchhoff's solution, but rather to obtain expressions that followed directly from the assumption of one of the two different boundary conditions on S-A. Kirchhoff's theory of course required that both be satisfied at once. Because Rayleigh had not fundamentally altered the physical presumptions that underpin Kirchhoff's (or Poincaré's) analysis, *viz* that the screen can annul the wave or its normal gradient, and that the wave in the aperture remains unaffected, his two possible integrals were distinguished solely by the *a priori* imposition of different conditions. That is, Rayleigh's theory, like Kirchhoff's, remained tied to imposed requirements without investigating what might underpin such stipulations on the grounds, just then becoming prevalent, of electromagnetic optics.

In the early 1890s the young German physicist Arnold Sommerfeld (1868–1951) developed a theory for diffraction that avoided Green's theorem by recurring to electromagnetic relations[55] He was at the time assistant to the mathematician Felix Klein (1849–1925) at the University of Göttingen. Like Kirchhoff and Poincaré, Sommerfeld was a dedicated teacher, and as a professor of theoretical physics at Munich from 1906 he trained an imposing array of students, including Wolfgang Pauli, Rudolf Peierls, Alfred Landé, Linus Pauling, I.I. Rabi and Max von Laue (the latter three as post-graduates).[56] In 1894 Sommerfeld, referring to Poincaré, had remarked in an article that Kirchhoff's boundary conditions are 'inadmissible' and would entail that wave function must vanish everywhere, though the theory nevertheless yields good agreement with observation (so far as was then known).[57] Two years later Sommerfeld published an intricate alternative that avoided the inconsistency but that was limited to the two-dimensional case of an infinitely thin, semi-infinite plane barrier with infinite electric conductivity struck by a linearly-polarized, plane wave parallel to it. To do so, Sommerfeld calculated the scattering of such a wave under the usual electromagnetic condition that the electric field vector in the plane of the screen vanishes, resulting after extensive, and intricate, calculations in an expression for the scattered wave.[58] The resulting expression indicated that a modified form of Young's original conviction that the illuminated edge of a diffractor could be treated as though it emitted waves. This result, however limited, constituted the foundation for a rigorous theory of diffraction, and to reach it Sommerfeld deployed a form of imaging in the complex plane in order to maintain the boundary condition on the electric field. Although the theory proved

---

[55]Sommerfeld 1896; translated into English as Sommerfeld 2004.

[56]For a comprehensive examination of the Sommerfeld school at Munich, and in particular its concentration on the solution of specific problems, see Seth 2010.

[57]Sommerfeld, 1895, pp. 341–342.

[58]In 1892 Poincaré had himself considered the case of diffraction from a sharp metallic edge at large diffraction angles in an effort to account for experimental results that Gouy had obtained (Gouy 1886). To do so he limited his analysis to infinite conductivity and examined the two-dimensional case by considering the wave from the edge produced by the scattering of a converging cylindrical disturbance whose axis parallels the edge (Poincaré 1892a; on this and Gouy's experiments see Darrigol 2015, pp. 14–16.) Sommerfeld referred to Poincaré's results for support since he had obtained a similar final expression under the same approximation (far field close to the edge of the geometric shadow): Sommerfeld 1896, p. 374.

influential over the years, the intricacy of the calculations continued to limit its application.[59] The use of an image procedure did nevertheless also lead Sommerfeld in his optical lectures to deploy images using Green's theorem proper. His approach there invoked the same duo of integrals that Rayleigh had used in 1897 albeit derived, unlike Rayleigh's, which worked with the form of the wave function, from a different specification of Green's function.[60]

Sommerfeld noted that if the Green's function $G(r)$ vanishes on S, then the term containing the normal derivative of the wave function will also vanish there:

$$\varphi(P) = \frac{1}{4\pi} \iint_S ds \left\{ \frac{\partial \varphi}{\partial N} G(r) - \varphi \frac{\partial G(r)}{\partial N} \right\},$$

$$G(r) = 0 \text{ over } S \Rightarrow \varphi(P) = \frac{-1}{4\pi} \iint_S ds \left\{ \varphi \frac{\partial G(r)}{\partial N} \right\}.$$

In that case only the values of the wave function and of the Green function's normal derivative need to be specified. Simple in principle, but difficult to apply in practice since there is no general Green's function that could satisfy such a requirement. Accordingly Sommerfeld had to limit his theory to a situation in which the Green's function could be so specified, and the only one that worked required the screen to be a plane.[61]

This can be done in the following way. Consider with Sommerfeld an infinite screen located at the origin and parallel to the xy plane. Place point P at (P_x, P_y, P_z) and consider an arbitrary point Q whose coordinates are (Q_x, Q_y, Q_z). Now reflect P across the screen to form its image P' whose coordinates will therefore be $(P_x, P_y, -P_z)$. Then the respective distances

---

[59]Sommerfeld produced a version of his theory in his lectures on optics: Sommerfeld, 1954 presented the lectures that he gave in 1934. Sections 38–39 provide his "mathematically rigorous solution" for the infinitely-thin, semi-infinite screen. See Born and Wolf 2002; chap. 11 develops the Sommerfeld theory up to the early 1950s. For subsequent developments see Babich, Lyalinov *et al.* 2007.
[60]Sommerfeld 1954, pp. 195–201. Sommerfeld had likely been giving these lectures for decades in some form, so that his version of the Rayleigh alternatives probably date to his early years at Munich, so a decade or more after Rayleigh's work on the subject. Sommerfeld did not mention Rayleigh in the published lectures, perhaps because his version of the alternatives involved a considerably different Green's function, as we shall see.
[61]*Ibid.*, 198–200.

r, r' from P, P' to Q will be:

$$r = \sqrt{(Q_x - P_x)^2 + (Q_y - P_y)^2 + (Q_z - P_z)^2},$$

$$r' = \sqrt{(Q_x - P_x)^2 + (Q_y - P_y)^2 + (Q_z + P_z)^2}.$$

Now define a new Green's function as follows:

$$G(Q) \equiv \frac{\exp(ikr)}{r} - \frac{\exp(ikr')}{r'}.$$

The new function is, as it were, formed by envisioning the *image* of a Green's function at point P across the screen and then summing the image with its progenitor. This new expression meets all the requirements for a Green's function, for it is a solution of Helmholtz's equation, converges to the appropriate form as r, r' reach zero, and adds nothing at infinity. We can calculate the derivative of G in the direction of the z-axis, *viz.* in the direction normal to the screen itself. Having done so, we move Q to the screen itself, where the result is that G vanishes. Over the shadowed part of the screen the wave function φ must vanish, while it must be continuous across the aperture. Sommerfeld assumed a point source and so expressed the emitted wave as $\frac{e^{ikr}}{r}$ where r is the distance from the source to a point of the aperture. For the alternative case in which the normal derivative of the Green's function must vanish, Sommerfeld merely had to form the *sum*

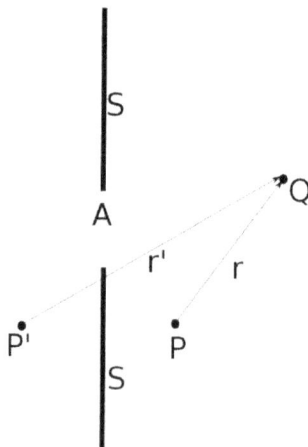

Figure 13:   Sommerfeld's configuration for his Green's function.

instead of the difference, *viz.* $G(Q) \equiv \frac{\exp(ikr)}{r} + \frac{\exp(ikr')}{r'}$, since the normal is opposite in direction on either side of the aperture.

Sommerfeld's route to a duo of mathematically consistent solutions differed considerably from Rayleigh's. Sommerfeld had altered the Green's function by forming a difference or sum between the original function and its image across the screen but had retained a spherical source wave. Rayleigh had instead retained the single-term Green's function but had required a plane source wave. Both proceeded to apply their expressions to the standard configuration in which, namely, the distance from the aperture to the field point is much greater than a wavelength. Moreover, though neither noted the point, the Kirchhoff integral is easily shown to be the *mean* of the Sommerfeld duo, which are limited to a plane screen, with Rayleigh additionally requiring the incident wave to parallel the screen.[62] Kirchhoff's full integral had no such limitation, but then it suffered from the Poincaré paradox.

$$\varphi^I(r) = \frac{1}{2\pi} \iint_A \varphi_{inc} \frac{\partial}{\partial N} \left(\frac{\exp(-ikr)}{r} \right) ds,$$

$$\varphi^{II}(r) = \frac{-1}{2\pi} \iint_A \frac{\exp(-ikr)}{r} \frac{\partial \phi_{inc}}{\partial N} ds,$$

$$\text{Kirchhoff: } \frac{1}{2} \left[\frac{1}{2\pi} \iint_A \varphi_{inc} \frac{\partial}{\partial N} \left(\frac{\exp(-ikr)}{r} \right) ds \right.$$

$$\left. - \frac{1}{2\pi} \iint_A \frac{\exp(-ikr)}{r} \frac{\partial \varphi_{inc}}{\partial N} ds \right].$$

The evaluation of either Rayleigh–Sommerfeld expression is much more difficult than for the Kirchhoff integral despite the limitation to a plane screen because neither of the duo individually yields the Fresnel integrals. In 1913, for example, Rayleigh applied φ^I to the case of an infinite narrow slit, which required extensive numerical computation via a series for the resulting integral.[63]

In his rigorous diffraction theory of 1896 Sommerfeld had concluded with a few remarks concerning the results of experiment that are worth

---

[62]For ϕ^I the normal derivative of the Sommerfeld difference Green's function reduces to twice the value of the positive term, while for ϕ^{II} the Sommerfeld function is just twice the value of either term.
[63]Rayleigh 1913.

quoting in full in order to gain a purchase on the empirical issues of the period. Since Poincaré in 1892 had obtained the same final results under the same circumstances, he concluded as follows:

> The question of the experimental confirmation of our [1896] theory can be here settled in a word. On the one hand Kirchhoff's formulas have been often proven by observations at a small diffraction angle; on the other hand Mr. Poincaré compares his formulas with Gouy's observations under large diffraction angles and finds them to be essentially confirmed. *In the same circumstances, that is for small as for large diffraction angle our theory therefore is also confirmed by experiment* [sic].[64]

Consequently Kirchhoff's full integral could account for experiments at small diffraction angles, but not at the large ones in Gouy's experiments, which latter could be handled by Poincaré's 1892 analysis based on electromagnetic relations (see note 58 above). Sommerfeld's scattering theory, which avoided Green theorem methods, could handle both.

In his optical lectures Sommerfeld did not directly address empirical issues, but he did point out that any theory involving the imposition of boundary conditions on an expression derived from Green's theorem (and so leading to an interpretation in terms of Huygens' principle) was questionable, including the Rayleigh–Sommerfeld duo:

> The question remains, are these assumptions also physically justifiable? The answer again is that *they are only approximations for sufficiently small wavelengths.* The field does not vanish completely behind the screen, nor is the field in the aperture entirely unaffected by the presence of the screen, at least not within distances of the order of magnitude of a wavelength from the edge of the screen. The introduction of the Green's function therefore involves no final justification of the method.[65]

Sommerfeld accordingly remained unconvinced that the boundary conditions associated with any such form of diffraction theory were physically reasonable, whereas the approach that he had developed in the early 1890s avoided the problem, at least to a certain extent. Nevertheless, half a decade after the appearance of Kirchhoff's paper an alternative expression based on his theory was produced that, we shall see, would lead decades later to a reformulation that could avoid the Poincaré paradox — but only at considerable mathematical cost.

---

[64]Sommerfeld 1896, p. 374.
[65]Sommerfeld 1954, p. 200. The italics are Sommerfeld's.

7. Kirchhoff's Integral Transformed

A year after Kirchhoff's death in the fall of 1887, and before the appearance of the Poincaré paradox, an Italian mathematician modified Kirchhoff's theory by converting it from a surface to a line integral, a transformation that implicated a different way to work with Kirchhoff's boundary conditions. A native of Milan, Italy, Gian Antonio Maggi (1856–1937) attended Kirchhoff's lectures in Berlin. In 1886 he had become an ordinary professor in analysis at the University of Messina.[66] Two years later Maggi published a transformation of Kirchhoff's fundamental integral that avoided the use of Kirchhoff's function F with its peculiar properties. Of his time in Berlin Maggi wrote in 1914 that he "was fortunate, not long before [writing his 1888 paper] to follow Kirchhoff's lessons at the University of Berlin, and so I was in a privileged position to pay attention to that supremely important result [Kirchhoff's diffraction theory], which the present state of the theory of the electromagnetic field has recently enriched with new applications."[67]

Kirchhoff was known for his strict adherence to precision and rigor. For example, the young Heinrich Hertz (1857–1894) — student of Kirchhoff's colleague Helmholtz — had sent him a paper on elastic collisions that Kirchhoff had extensively edited because Hertz had not been sufficiently explicit in laying out the precise conditions of the problem.[68] Perhaps Kirchhoff had expressed to Maggi directly or in his lectures some doubts concerning the propriety of his function F, or it may be that Maggi noted its presence and decided to avoid it in order to purify Kirchhoff's integral of one possible objection.[69] To do so Maggi deployed three points: two of the points are fixed, while the third is not. The wave equation's solutions V have the following usual form, with r representing the distance between a fixed point x_0, y_0, z_0 and a point x, y, z within the

---

[66]On Maggi see Cisotti 1938, Anonymous, last access 15 July 2015 (Maggi 1888). The reminiscence is from Maggi, 1914. Maggi wrote the latter paper in order to compensate the difficulty of his 1888 presentation which, he wrote in 1914, "had somewhat harmed its perspicuity."

[67]Maggi 1914. Since Kirchhoff died in 1887, Maggi's "not long before" means several years prior to 1888.

[68]On which see Buchwald 1994, chap. 8.

[69]The fact that the addition by Hensel in the published lectures a few years later writes of F that such a function "actually" exists may indicate doubt at the time concerning the propriety of basing such a fundamental result upon it.

enclosed volume:

$$\frac{\partial^2 V}{\partial t^2} = a^2 \nabla^2 V \quad \text{with } r = \sqrt{(x_0 - x)^2 + (y_0 - y)^2 + (z_0 - z)^2}$$

$$V = \frac{\varphi(t - r/a)}{r}.$$

Maggi then noted that the function can contain, separately, the variable coordinates x, y, z and still satisfy the wave equation, so that V becomes $V(t - r/a, x, y, z)$. By considering the two fixed points, one of whose coordinates (x, y, z) appear in the expression for r whereas the coordinates of the other does not, Maggi was able to obtain Kirchhoff's fundamental integral without using the function F while nevertheless leaving the form of the wave open. The reasoning was sufficiently obscure that he attempted a simplification in 1914 (see above, note 66).

This was not Maggi's only innovation. In addition, he effected a significant transformation of the fundamental integral. To do so he used the line joining his two fixed points to specify the edge of a plane that could be used in a complicated manner to create a surface that separated space into two contiguous regions within which the analysis took different forms. These two points represented the loci of the source and of the point at which the wave is observed past a surface that represents a screen, and the regions in question correspond respectively to the space defined by the geometric shadow and to the space outside it. Kirchhoff's boundary conditions were separately applied to these two regions. The result of the procedure yielded a complicated function (due to the choice of prolate spheroidal coordinates) for the value of the wave that involved in both regions a line integral around the edge of an aperture. In the geometrically-illuminated region the disturbance is determined by the sum of a direct wave from the source added to the line integral, while in the geometric shadow only the line integral holds. This accordingly implicated a *discontinuity* across the shadow boundary, though Maggi's use of prolate coordinates rather obscures the configuration. The transformation to a line integral was effected by Maggi's use of Stokes' theorem based on the assumption that the medium is incompressible, as was usual. If, then, $\vec{u}$ represents a disturbance in such a medium, $\nabla \cdot \vec{u}$ must vanish, in which case $\vec{u}$ can be represented by $\nabla \times \vec{v}$ with $\vec{v}$ an appropriate function. Consequently Stokes' theorem affords the transformation of a surface integral for $\vec{u}$ into a line integral for $\vec{v}$:

$$\iint_\sigma \vec{u} \cdot d\sigma = \iint_\sigma (\nabla \times \vec{v}) \cdot d\sigma = \oint_{\partial\sigma} \vec{v} \cdot d\vec{l}.$$

In a brief consideration at the end of his article, Maggi concluded that, in the limit of vanishingly small wavelength, the disturbance occurs entirely outside the region of the geometric shadow. The signal purpose of his transformation to a line integral, then, was to yield that result, which in effect retrieved geometric optics. Yet it seems that very few people paid attention to this latter result, perhaps because it was obscurely framed in terms of prolate coordinates. Which makes it hardly surprising that the first to re-achieve such a transformation did not mention Maggi at all.

Wojciech (Adalbert) Rubinowicz (1889–1970) was born in Bukovina to Polish parents, and received his Ph.D. in physics at the University of Czernowitz in 1916. He planned to stay there as a post-graduate assistant, but the university closed during the First World War. He obtained a temporary position at the University of Munich's institute of theoretical physics, where he began work as assistant to Arnold Sommerfeld in 1916.[70] A year later, Rubinowicz published an article in the *Annalen der Physik* that explored a line-integral expression for Kirchhoff's fundamental integral.[71] He was stimulated to do so by Sommerfeld's 1894 analysis of diffraction based on electromagnetic theory (on which see above), as Rubinowicz aimed to see whether a similar result could be obtained directly from Kirchhoff's integral by means of a line-integral transformation. Although Rubinowicz's work became much better known among European physicists and mathematicians than Maggi's, the conversion that both effected, though in considerably different ways, would later become known as the "Maggi–Rubinowicz transformation."[72]

Rubinowicz did not maintain Kirchhoff's original formulation that allowed for an arbitrary wave but instead used the reduced Helmholtz equation and so presumed that the time and space variables could be separated. Consequently, unlike Maggi he did not have to concern himself with Kirchhoff's delta-like function. Rubinowicz first introduced the general surface integral over the boundary of a region G with a Green's function e^{ikr}/r, with $\bar{u}$ representing the value of the wave function over the region's boundary:

$$\frac{1}{4\pi} \iint_G ds \left\{ \bar{u} \frac{\partial(e^{ikr}/r)}{\partial n} - \frac{e^{ikr}}{r} \frac{\partial \bar{u}}{\partial n} \right\}. \tag{18}$$

---

[70] Eckert 2013, p. 226.

[71] Rubinowicz 1917.

[72] In 1923, the Austrian professor of physics Friedrich Kottler at the University of Vienna pointed out that Maggi, not Rubinowicz, had first produced the transformation. See Kottler 1923.

He then remarked that Kirchhoff had reduced expression (18) to an integral solely over an aperture F, producing the following expression for the wave at a point O in the diffraction region:[73]

$$u(O) \equiv \frac{1}{4\pi} \iint_F ds \left\{ \bar{u} \frac{\partial(e^{ikr}/r)}{\partial n} - \frac{e^{ikr}}{r} \frac{\partial \bar{u}}{\partial n} \right\}.$$

Rubinowicz next effected a clever choice of region and geometric boundary for the otherwise arbitrary surface G, one that was effectively the same as Maggi's but considerably clearer due to the avoidance of prolate coordinates. Limit G to a surface consisting of the region that would be geometrically illuminated by a source L of the form $\exp(ik\rho)/\rho$ shining through an aperture F which has a rim B. This demarcates a region with one end a surface K_∞ at infinity, capped at the other by the finite aperture F, and delimited by the conical boundary K (Figure 14). The integral over K_∞ vanishes as usual, leaving only the region $F + K$. Since we have not introduced a physical aperture, the value of $\bar{u}$ over $F + K$ must be the same that it would be in the absence of the screen, hence just $\frac{\exp(ik\rho)}{\rho}$. Consequently the expression for the field at a point O_E within the

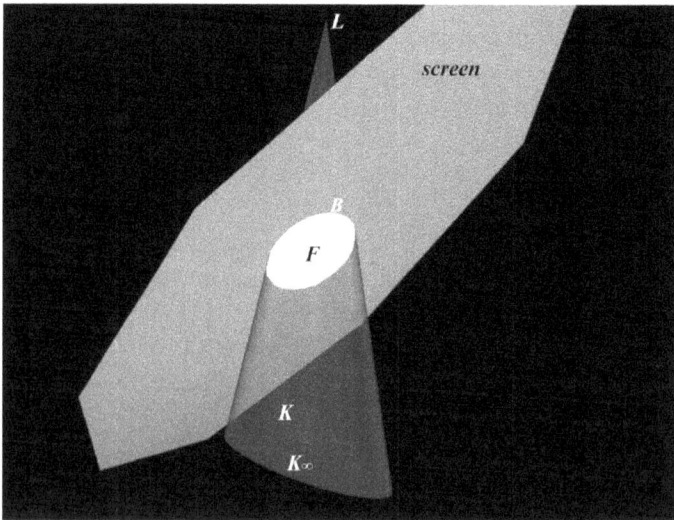

Figure 14:   Rubinowicz's surfaces.

---

[73]Rubinowicz 1917, p. 259.

$F + K$ bounded region — which is, again, completely unobstructed — must be:[74]

$$u_E(O_E) \equiv \frac{1}{4\pi} \iint_{F+K} ds \left\{ \frac{\exp(ik\rho)}{\rho} \frac{\partial}{\partial n} \left(\frac{\exp(ikr)}{r} \right) \right.$$

$$\left. - \frac{\partial}{\partial n} \left(\frac{\exp(ik\rho)}{\rho} \right) \left(\frac{\exp(ikr)}{r} \right) \right\} = \frac{\exp(ik\rho)}{\rho}. \qquad (19)$$

Introduce a screen such that the surface F corresponds to a physical aperture. In that case for a point O located anywhere in the diffraction region Kirchhoff's integral requires (with $\bar{u}$ on F having the same value that it would were the screen absent):[75]

$$u(O) \equiv \frac{1}{4\pi} \iint_F ds \left\{ \frac{\exp(ik\rho)}{\rho} \frac{\partial}{\partial n} \left(\frac{\exp(ikr)}{r} \right) \right.$$

$$\left. - \frac{\partial}{\partial n} \left(\frac{\exp(ik\rho)}{\rho} \right) \left(\frac{\exp(ikr)}{r} \right) \right\}. \qquad (20)$$

Now place the general Kirchhoff point O at O_E within the region bounded by $F + K$. Then for such a point, since equation (20) integrates over F only, whereas equation (19) integrates over $F + K$, Rubinowicz could subtract equation (20) from equation (19) to write:

$$u(O_E) = u_E(O_E) - \frac{1}{4\pi} \iint_K ds \left\{ \frac{\exp(ik\rho)}{\rho} \frac{\partial}{\partial n} \left(\frac{\exp(ikr)}{r} \right) \right.$$

$$\left. - \frac{\partial}{\partial n} \left(\frac{\exp(ik\rho)}{\rho} \right) \left(\frac{\exp(ikr)}{r} \right) \right\}. \qquad (21)$$

Rubinowicz's expression (21) for $u(O_E)$ applies only to a point within the $F + K$ bounded region. Outside of this region Kirchhoff's general expression (20) for $u(O)$ would presumably apply. This might seem to be a pointless formality — we have as it were introduced surface K in equation (19) only to remove it again in equation (21). But we have in fact done something more, because we assert that equation (21) does properly represent the resultant wave within the region bounded by $F + K$ when a screen perforated by F

---

[74] *Ibid.*, p. 260.
[75] *Ibid.*, p. 259.

is present. This is nothing more than Kirchhoff's claim for such a region but Rubinowicz's equation (21) has an important representational effect.

Because the value of the normal gradient $\frac{\partial \bar{u}}{\partial n}$ vanishes over K, Rubinowicz could transform the K integral into one entirely around the rim B of the cap F. Integrating from $\rho = \rho_l$ (at a point on the rim) to infinity thereby produced (recall that ρ in the K integral is the distance between L and a point on K):

$$\iint_K ds = \oint_B dl \int_{\rho_l}^{\infty} d\rho. \tag{22}$$

Attention to the geometry yields an expression for the wave field within $F + K$ that involves an integral over the rim B added to the geometric-optics field:

$$u(O_E) \equiv u_E(O_E) + \frac{1}{4\pi} \oint_B dl \left\{ \frac{\exp[ik(\rho_l + r_l)]}{\rho_l r_l} \frac{\cos(\hat{n}, \vec{r}_l)}{1 + \cos(\vec{r}_l, \vec{\rho}_l)} \sin(\vec{\rho}_l, d\vec{l}) \right\}. \tag{23}$$

In Rubinowicz's equation (23), r_l is the distance between the field point O and a point l on B, $\hat{n}$ is the normal there to the screen aperture F, $\vec{\rho}_l$ is the distance from the rim point to the luminous point at L, and $\cos(\hat{n}, \vec{r}_l)$ is the cosine of the angle between the two vectors $\hat{n}$ and $\vec{r}_l$. The source wave $u_E(B)$ at the rim appears as the factor $\frac{\exp[ik\rho_l]}{\rho_l}$, so that equation (23) may also be written as follows:

$$u(O_E) \equiv u_E(O_E) + \frac{1}{4\pi} \oint_B dl \left\{ u_E(B) \frac{\exp[ikr_l]}{r_l} \frac{\cos(\hat{n}, \vec{r}_l)}{1 + \cos(\vec{r}_l, \vec{\rho}_l)} \sin(\vec{\rho}_l, d\vec{l}) \right\}. \tag{24}$$

Rubinowicz's equation (24) entails an inherent discontinuity at the boundary K: on it the value of the rim integral vanishes because there the two vectors $\vec{r}_l, \vec{\rho}_l$ point in opposite directions, leaving only the unobstructed geometric field $u(O_E)$. In the region external to $F + K$ the untransformed Kirchhoff expression (20) is presumably obtains, and so also do his conditions that both the wave function and its normal gradient vanish over the shadowed surface of the screen. Consequently, and despite the fact that Kirchhoff's $u(O)$ still holds everywhere, the Maggi–Rubinowicz introduction of the $F + K$ boundary introduces a representational discontinuity, though one that neither of the two explored since their interests lay elsewhere. Neither did they ask whether the rim integral would vanish if the observation point O_E lay in the aperture proper, where on the basis of

Kirchhoff's boundary condition — which is still presumed — the incident wave $u_E(O_E)$ should be unaltered.

Although Rubinowicz's equation (24) was in essence similar to Maggi's transformation, Maggi's use of prolate coordinates obscured a startling result upon which Rubinowicz placed considerable emphasis (and which Maggi himself likely did not perceive). Namely, that equation (24) amounted to a partial vindication of Thomas Young's original method of calculating diffraction by finding the interference between a direct wave from the source and waves presumptively engendered by the direct wave at the edges of the aperture — as Rubinowicz remarked at the very beginning of his article.[76] This amounted to a considerable alteration in the original sense of the Huygens — Fresnel principle from one in which the observed wave is governed entirely by wavelets that are distributed over the aperture's surface, to one in which the efficacious wavelets that alter the incident wave are located on the aperture's rim.

Rubinowicz's article appeared in 1917, over three decades after Kirchhoff's original and Maggi's own transformation, and neither he nor Maggi had concerned themselves with the problem of Kirchhoff's boundary conditions (which Maggi may not have been aware of in 1888). Each of them had a different purpose in mind, Maggi to avoid Kirchhoff's function F, which led him to his line-integral transformation, and Rubinowicz to retrieve a simulacrum of Young's original theory of diffraction. Does the *Poincaré paradox* carry over as well to the Maggi–Rubinowicz transformation? Like Maggi, but with greater geometric clarity since he had not introduced the complexity of prolate coordinates, Rubinowicz had altered the way in which Kirchhoff's boundary conditions had been applied by both Kirchhoff and by Poincaré by introducing two distinct regions across which the expression for the wave function is discontinuous. Neither of these regions had a boundary coincident with the entire surface of the diffractor proper since the one was capped by the aperture and the other by the surface outside the geometric shadow. For that reason Poincaré's proof of inconsistency, which required no discontinuities as well as the enclosing surface of the diffractor (within which Poincaré placed his point L to generate the paradox), could not be so directly applied. The question accordingly arises as to whether the new

---

[76] Partial only because Young had assumed that the wave emanating from the aperture's rim would be the same as the source wave but shifted in phase by half a wavelength. Rubinowicz's rim integral is considerably different.

representation might capitalize on discontinuity to evade the Poincaré paradox. To do so in a way that preserved the Green theorem structure in some form would certainly require altering Kirchhoff's boundary conditions.

8.   A Form of Consistency Achieved

Work continued in the twentieth century through at least two alternatives that, though not always aimed principally at that purpose, could avoid the Poincaré paradox. In 1923 Friedrich Kottler (1886–1965), physics professor at the University of Vienna, proposed formulating the integral as a solution to a "saltus" problem — one in which the field undergoes a discontinuous jump across a boundary, requiring the specification of its values on either immediate side, much as, in electrostatics, a fictitious distribution of charge at the boundary across which dielectric capacity changes produces a discontinuity in the potential.[77] Kottler proposed to treat Kirchhoff's conditions as though they were the result of a similar kind of discontinuity, thereby obviating the Poincaré paradox but at the cost of introducing an assumption without physical justification.[78] In 1933 Max Born (1882–1970), thinking in terms of his characteristic recursive method, suggested that the Kirchhoff integral might be the first-order approximation in a sequence of iterative solutions that converged to the exact solution represented by his boundary conditions.[79]

In 1964 Marchand and Wolf remarked that "the difference between the consistent solution of Rayleigh and Sommerfeld and the inconsistent solution of Kirchhoff may be regarded as due to the superposition of plane waves whose amplitude distribution has a very sharp maximum for . . . waves propagated along the plane of the screen. Such waves do not contribute to the far field."[80] Two years later they followed with an analysis that enabled a direct calculation of precisely such a series of waves by means of a reworking of the boundary conditions through a clever use of the Maggi–Rabinowicz transformation, thereby rescuing the essence of Kirchhoff's formulation and, with it, the meaning of an expression akin to Huygens principle.

---

[77] Kottler, 1923 and, decades later, Kottler 1965.

[78] By contrast, in electro- and magneto-statics dielectric and para- or dia-magnetic substances yield the discontinuity as a result of the jump in permeability at a boundary.

[79] Born 1933, p. 152.

[80] Marchand & Wolf, 1964.

Rubinowicz's form of the transformation, recall, could be expressed in terms of the source as follows (*cf* equation (23)) invoking a discontinuity for points past the screen as one moves from within to without the geometric shadow since the denominator in the formula vanishes for points on the shadow proper:

$$u(O_E) \equiv u_E(O_E) + \frac{1}{4\pi} \oint_B dl \left\{ u_E(B) \frac{\exp[ikr_l]}{r_l} \frac{\cos(\hat{n}, \vec{r}_l)}{1 + \cos(\vec{r}_l, \vec{\rho}_l)} \sin(\vec{\rho}_l, \vec{dl}) \right\}.$$

Rubinowciz's transformation of Kirchhoff's integral within the geometrically illuminated region.

Kirchhoff's surface-integral involves no such discontinuity of the kind for it leads to the following expression (*cf* above, and note 41):

$$\varphi_o = \frac{1}{2\lambda} \iint_A \frac{ds}{r_i r_o} \left[\cos(\vec{r}_i, \vec{n}_i) + \cos(\vec{r}_o, \vec{n}_o) \right] \sin \left[2\pi \left(\frac{r_i + r_o}{\lambda} - \frac{t}{T} \right) \right].$$

Marchand and Wolf considered that the inconsistency built into Kirchhoff's theory by his boundary conditions could be shown in a different manner from Poincaré by noting a related consequence. Take a point in the diffracted region (not, as with Poincaré, within the geometrically-defined body of the diffractor itself) and move it indefinitely close to the open aperture. Calculate the value of the wave function at that point using the expression that results from Kirchhoff's theory. It should by hypothesis be effectively the same as the unaltered source wave, but in fact it is not.[81] Why, one might ask, did they turn to this way to show the inconsistency? It is quite direct and perhaps more revealing than Poincaré's route, but, more to the point, it is precisely this consequence of the

---

[81] The expression given immediately above is not applicable to a point close to the aperture because of the assumption that the aperture point to observation point distance is vastly larger than a wavelength. If however we take as an example of the problem what occurs in the case of a circular disk as screen and make no approximations, then two problems arise using the full Kirchhoff integral. The solution along the axis produces two terms, the second of which is infinite and obviously unphysical. Moreover — and this is the sort of problem that Marchand and Wolf had in mind — even the first term does not vanish at the disk itself, where by hypothesis no wave at all should exist. In other words, the boundary condition presupposed cannot be recovered (see Lucke 2004, pp. 3–4). A similar situation arises for diffraction by a circular aperture, in which the incident wave is not recovered at the aperture.

full set of Kirchhoff's conditions that can be avoided by exploiting the
discontinuity at the geometric shadow inherent in the Maggi–Rubinowicz
transformation.[82]

To that end, Marchand and Wolf began not with the transformation
but with an explicit alteration of the boundary conditions, something that
neither Maggi nor Rubinowicz had proposed. Consider with Marchand and
Wolf a luminous point L whose distance from a given point of the aperture
rim is s_0. Consider a point P' with coordinate (x, y) on the shadowed part
of the screen itself, including any aperture(s), and express the presumptive
value $U(P')$ of the wave at such a point of the screen by a sum of the source
wave added to an integral over the aperture's rim B, of the following form,
with points Q' lying on the rim:

$$U(P') \equiv f_0(x,y) = \varepsilon_0(x,y)u_{inc}(x,y) + \frac{1}{4\pi} \oint_B u_{inc}(Q') \frac{\exp(ikQ'P')}{Q'P'} \vec{K} \cdot \vec{dl}$$

where

$$\vec{K} = \frac{1}{4\pi} \frac{\overrightarrow{SQ'} \times \overrightarrow{Q'P'}}{\left|\overrightarrow{SQ'}\right|\left|\overrightarrow{Q'P'}\right| - \overrightarrow{SQ'} \cdot \overrightarrow{Q'P'}}$$

and where

$\varepsilon_0 = 0$ for points on the screen but outside the aperture;

$\varepsilon_0 = 1$ for points in the aperture proper;

S is the location of a point light source outside the aperture.

This expression for the field intensity $U(P')$ on both the screen's shadowed surface and on the aperture constitute *new boundary conditions* for
Kirchhoff's problem of diffraction. Where Kirchhoff had set the value at the
aperture to the incident wave, Marchand and Wolf added to that wave the
rim integral. And where Kirchhoff had set the wave on the screen but outside the aperture to zero, they now added the rim integral. Neither Maggi
nor Rubinowicz had suggested that the rim integral might be extended to
points of the screen outside the aperture, and neither had they considered
what takes place if the observation point were to lie in the aperture proper,
where, again, on Kirchhoff's boundary condition the rim integral should
vanish to retrieve the unaltered incident wave.

In a critical next step, Marchand and Wolf note that their expression for
$\vec{K}$ is identical to the one that appears in the Maggi–Rubinowicz transformation of Kirchhoff's diffraction integral, i.e. in the factor of $u_E(B)\frac{\exp[ikr_l]}{r_l}$

---

[82]Marchand & Wolf 1966.

in Rubinowicz's rim integral (24). From here, they claim that the diffracted wave field at a point $P = (x, y, z)$ on the side of the aperture opposite to the light source $(z > 0)$ can be expressed as

$$U_K(x, y, z) = \varepsilon_0(x, y, z)u_{inc}(x, y, z) + \frac{1}{4\pi} \oint_B u_{inc}(Q') \frac{\exp(ikQ'P)}{Q'P} \vec{K} \cdot d\vec{l}.$$

Although the Marchand–Wolf integral appears to differ from the term that appears in Rubinowicz's expression of Kirchhoff's formula (24), they are nevertheless essentially equivalent and so both exhibit precisely the same discontinuity: at the geometric shadow proper, i.e. if P' lies on K, $\overrightarrow{SQ'}$ parallels $\overrightarrow{Q'P'}$, and consequently the expression for $\vec{K}$ is discontinuous there. Now, they continued, the rim integral alone should be extended to *any point* outside the geometrically-illuminated region. How so, when Rubinowicz's derivation apparently concerned only the latter? Though Marchand and Wolf did not explicitly provide the reason, it is not at all hard to see: if Kirchhoff's boundary condition outside the geometric region but on the screen is replaced by Marchand–Wolf's (with $\varepsilon_0 = 0$), then on the conical sides delimiting the region the normal derivative of the wave function again vanishes, while the aperture rim still delimits the region, leaving, via the Maggi–Rubinowicz transformation, only the rim integral over the closed part of the screen since the latter is bounded by the aperture.

Consequently the Marchand–Wolf expression for $U(P')$ now holds throughout the diffraction region (in the form of $U_K(x, y, z)$ above) provided that ε_0 is unity everywhere within the geometrically-illuminated region but vanishes outside it. Moreover, because of the new condition at the aperture, the Marchand–Wolf integral for a point in the diffraction region that is indefinitely close to the aperture now reproduces the presumed value there (i.e., $U_K(x, y, z = 0) \to f_0(x, y)$). This is precisely what does not happen when using Kirchhoff's original with his condition that the wave at the aperture is unaltered from what it would be in the screen's absence. And this is all obtained from the transformation of Kirchhoff's original integral combined with Marchand and Wolf's alteration of the boundary conditions. The Kirchhoff inconsistency is thereby avoided altogether since the new boundary conditions involve only the value of the wave function proper and not is normal gradient, whether over the screen proper or over the aperture — at the expense of discontinuity at the geometric shadow edge.

That is not the only cost, for the transformation to a line integral also implicates a significant amendment to the deployment of Fresnel's integrals, which had (and continue to this day) to permeate practical computations of diffraction since the 1830s. Within the approximation regime that governed the Fresnel (and so Kirchhoff) expressions, namely small wavelengths and loci far from the screen but not too far from the edge of the geometric shadow, these original expressions work well despite the boundary inconsistencies necessary to their derivation. But, one might ask, does the Marchand–Wolf replacement do any better? Indeed, their 'consistent' version of the Kirchhoff integral does, they showed, produce good agreement with an experiment in which the behavior of 3.2 cm microwaves was examined across the aperture in a conducting screen. The experiment indicated that the wave pattern across it varies nearly sinusoidally, which is well predicted by Marchand–Wolf but not of course by Kirchhoff's theory, which requires the unaltered value of the source wave.[83] The Marchand–Wolf expression explains the results as due to the interference of the source wave with the wave scattered at the aperture's edge. But the sinusoidal behavior in the aperture nevertheless does imply that the Kirchhoff assumption of an unaltered incident wave at the aperture works quite well for points in the diffraction region that are not too close. This retrieves and justifies the Kirchhoff integral within an approximation regime with its longstanding, and persuasive, interpretation in terms of Huygensian wavelets.

9. Conclusion

We have examined a diffraction theory that, for nearly two centuries (including Fresnel's original) continued in use even though, a century in, its only derivation from a wave equation had nevertheless proved to be fundamentally inconsistent with the very conditions that yielded the expression that, so far as was then known, worked extremely well. One major reason for that resilience was the theory's instantiation within a mathematical framework of the principle, due to Huygens, on which Fresnel had originally grounded wave optics. Whereas Fresnel's wave optics had deployed assumptions concerning point-source radiators that presumptively constitute a wave front, Kirchhoff's theory was formulated as a solution to a

---

[83]Neither are these results accommodated by the Rayleigh–Sommerfeld alternative in which the waveform at the aperture is assumed to be the same as that of the incident disturbance. The second alternative, in which the waveform's normal derivative is specified, fares somewhat better but still misses the mark. *Ibid.,* pp. 1715–1717.

partial differential equation based on Green's theorem. Critically, so far as was known the resulting formula for diffraction worked extremely well.

Empirical adequacy and pragmatic tractability certainly were — and remain — central factors for the tenacity of Kirchhoff's theory. Scientists have, and continue to this day, to use the expression provided by Kirchhoff's theory — or even Young's much simpler structure involving two-ray interference — in astronomical observations and optical experiments. For example, in Albert Michelson and his colleague F.G. Pease's interferometric measurements at Mt. Wilson Observatory as late as 1920, the wave-optical method they employed to estimate a star's diameter did not go beyond the Fresnel integrals proper.[84] Throughout the late nineteenth and twentieth centuries, Kirchhoff's theory continued to appear in major textbooks and monographs, continued to be used and discussed in physics and engineering periodicals, and was generally considered to be a reasonable expression for the effect of diffraction by an aperture. The problem of consistency, to the extent — and it was apparently not a very great extent — that it was recognized seemed principally to concern a defect in the manner in which the Huygens principle or some variant thereof could be represented within a consistent mathematical context.

The principal alternative to Kirchhoff's theory involved dropping one or the other of the two mutually-inconsistent boundary conditions. Rayleigh, who was not concerned with the Kirchhoff problem *per se*, had produced both in 1896 by choosing particular wave functions, while Sommerfeld elaborated the two alternatives using novel Green's functions, with the consequent result that Kirchhoff's solution proved to be the mean of the Rayleigh–Sommerfeld duo, though without any clear reason as to why that might be so. When Marchand and Wolf compared newly-produced microwave data with predictions from Kirchhoff's and the Rayleigh–Sommerfeld expressions in 1966, they found that Kirchhoff's, inconsistent mathematically though it certainly is, better matched measurements at the aperture.[85] To a certain extent such a result was not altogether surprising since Sommerfeld did not regard the boundary conditions that he had himself used to be physically reasonable, even if mathematically unexceptionable.[86] By contrast, Sommerfeld's own analytical solution in 1896, based on electromagnetic relationships, though limited to diffraction by the

---

[84]Michelson & Pease 1920.
[85]Marchand & Wolf 1966, pp. 1715–1717.
[86]Sommerfeld 1954, p. 200.

edge of a half-infinite plane, did not require any *a priori* assumptions concerning boundary conditions. Yet the extreme complexity and intractability of this approach significantly restricted its practical applicability, whereas Kirchhoff's integral easily produced an expression under the approximation to small wavelengths in relation to aperture size and distance from the screen for both source and observation point that easily facilitated computations and was widely used in optical research. In that approximation Kirchhoff's theory was essentially a differential-equation-based reformulation of Fresnel's formula, a major theoretical result in early wave optics that had enjoyed significant empirical success.

The inconsistency problem was never, in one sense, truly resolved because it follows directly from the requirements of Green's theorem. And yet in the absence of Green's theorem a comparatively direct reading in terms of anything like Huygens' principle, with its persuasive physical significance, evaporated. Kottler's recommendation to reformulate the theory as a *saltus* problem amounted to introducing a fundamental *discontinuity* in applying Green's theorem in a manner that could retain the interpretative meaning of Huygens' principle. Born's suggestion to interpret the Kirchhoff expression in terms of a first-order expansion of whatever the actual solution might be was not directly aimed at preserving the physical significance granted to Huygens' principle. It was nevertheless also an effort to reconstruct Kirchhoff's mathematical structure in a manner that separated his solution, which certainly did grant meaning to the principle, from its grounding in a mathematical inconsistency.

In 1964 Marchand and Wolf employed the long-available Maggi–Rubinowicz transformation to replace Kirchhoff's single surface scheme with one in which the solution is applied separately to two contiguous regions across which the expression for the wave function is discontinuous. That breaks the inconsistency otherwise demanded by Dirichlet–Neumann solution requirements and accordingly breaches the proof of the Poincaré paradox. The resulting expression aligned well with Young's theory in the early nineteenth century, in which light rays from an optical source reach points within the geometrically-illuminated region directly and from the aperture's rim, and outside the region solely from the rim. This new physical meaning, gained through discontinuity and transformation to a line-integral, made the scheme particularly appealing — but not for computational purposes under the usual approximation to vanishingly small wavelength, for there the Fresnel integrals work well for reasons that, in light of Marchand–Wolf, have become clear. The price to be paid was the abandonment of Kirchhoff's original, extremely simple, boundary conditions over the surface of the screen via the introduction of discontinuity.

The history of Kirchhoff's solution illustrates that mathematical consistency is not inevitably a necessary condition for success in physics. While logical compatibility, mathematical rigor, and conceptual coherence are important epistemic virtues for theory choice and development, they are not inexorably requisite. Empiricist and pragmatic attitudes prompted optical scientists significantly to downplay the mathematical inconsistency of Kirchhoff's theory, a theory that exemplifies a long research program that dominated optics for two centuries. Rather than abandoning altogether such a useful scheme because of mathematical requirements, researchers instead preferred to use it and eventually to reconstruct the mathematics on a different, if doubtfully rigorous (because involving discontinuity) basis, and thereby to grant it continued life.

Bibliography

Anonymous. (last access 15 July 2015) Maggi, Gian Antonio, *Treccani Enciclopedia Italiana*, from http://www.treccani.it/enciclopedia/gian-antonio-maggi/.

Babich, V. M., M. A. Lyalinov and V. E. Grikurov (2007). *Diffraction Theory. The Sommerfeld-Malyuzhinets Technique*, St. Petersburg: Alpha Science Intl. Ltd.

Baker, B. and E. T. Copson (1939) *The Mathematical Theory of Huygens' Principle*, Oxford: Clarendon Press.

Bloor, David (2011) *The Enigma of the Aerofoil: Rival Theories in Aerodynamics, 1909–1930*, Chicago: University of Chicago Press.

Born, Max (1933) *Optik*. Berlin: Julius Springer-Verlag.

Born, Max and Emil Wolf (2002) *Principles of Optics*, Cambridge: Cambridge University Press.

Buchwald, Jed Z. (1980) Optics and the Theory of the Punctiform Ether, *Archive for History of Exact Sciences* 21: 245–278.

——— (1981) The Quantitative Ether in the First Half of the Nineteenth Century. *Conceptions of Ether: Studies in the History of Ether Theories, 1740–1900*, ed. by G. Cantor and M. J. S. Hodge. Cambridge, Cambridge University Press: 215–237.

——— (1985) *From Maxwell to Microphysics. Aspects of Electromagnetic Theory in the Last Quarter of the Nineteenth Century*, Chicago: University of Chicago Press.

——— (1989) *The Rise of the Wave Theory of Light. Optical Theory and Experiment in the Early Nineteenth Century*, Chicago, University of Chicago Press.

———(1994) *The Creation of Scientific Effects. Heinrich Hertz and Electric Waves*, Chicago: University of Chicago Press.

———(2012) Cauchy's theory of dispersion anticipated by Fresnel. *A Master of Science History*, ed. by Jed Z. Buchwald. Dordrecht: Springer. Chapter 20: 399–416.

———(2013) Optics in the Nineteenth Century, in: *The Oxford Handbook of the History of Physics*, ed. by J. Z. Buchwald and R. Fox. Oxford: Oxford University Press: 445–472.

Charpentier, E., E. Ghys and A. Lesne, Eds. (2010) *The Scientific Legacy of Poincaré*. Providence, RI: American Mathematical Society.

Cheng, A. H. D. and D. T. Cheng (2005) Heritage and early history of the boundary element method, *Engineering Analysis with Boundary Elements* 29: 268–302.

Cisotti, U. (1938) Gli scritti scientifici di Gian Antonio Maggi, *Rendiconti del Seminario Matematico e Fisico di Milano* 12: 167–189.

Cross, J. J. (1985) Integral theorems in Cambridge mathematical physics, 1830–1855, in: *Wranglers and Physicists. Studies on Cambridge Mathematical Physics in the Nineteenth Century*. ed. by P. M. Harman. Manchester: Manchester University Press: 112–148.

Darrigol, Olivier (2012) *A History of Optics from Greek Antiquity to the Nineteenth Century*. Oxford: Oxford University Press.

———(2015) Poincaré's Light. *Henri Poincaré, 1912–2012: Poincaré Seminar 2012*, ed. by B. Duplantier and V. Rivasseau. Basel: Springer: 1–50.

Dieudonné, Jean (2008) Jules Hernri Poincaré. *Complete Dictionary of Scientific Biography*. Detroit, Charles Scribner's Sons: 59–60.

Eckert, Michael (2013) *Arnold Sommerfeld: Science, Life and Turbulent Times, 1868–1951*, New York, Springer.

Goodman, J. W. (1988) *Introduction to Fourier Optics*, New York: McGraw Hill.

Gouy, L. G. (1886) Recherches experimentales sur la diffraction, *Annales de chimie et de physique* **52**: 145–192.

Gray, Jeremy (2013) *Henri Poincaré: A Scientific Biography*, Princeton: Princeton University Press.

Green, George (1828) An Essay on the Application of Mathematical Analysis to the Theories of Electricity and Magnetism. *Mathematical Papers of the Late George Green*, Cambridge: Cambridge University Press: 1–82.

Hentschel, Ann (2016) Gustav Kirchhoff's treatise *On the theory of light rays*, in K. Hentschel and N. Zhu (eds.) *Gustav Robert Kirchhoff's Treatise On the Theory of Light Rays (1882) — English Translation, Analysis and Commentary*. Berlin, Logos-Verlag: (Stuttgarter Beiträge zur Wissenschafts-u. Technikgeschichte, 10), here on pp. 43–76.

Herschel, John (1827) Light, in: *Encyclopedia Metropolitana*, completed in 1827, publ. London, 1845, vol. 9: 341–586.

Jungnickel, Christa and Russell McCormmach (1986) *Intellectual Mastery of Nature: Theoretical Physics from Ohm to Einstein*, Chicago: University of Chicago Press.

Kipnis, Nahum (1991) *History of the Principle of Interference of Light*, Basel: Birkhäuser.

Kirchhoff, Gustav Robert (1876) *Vorlesungen über Mathematische Physik, Bd 1: Mechanik*. Leipzig: B. G. Teubner.

―――(1882) Zur Theorie der Lichtstrahlen, *Sitzungsberichte der Königlich Preussischen Akademie der Wissenschaften zu Berlin*, part 2: 641–669.

―――(1883) Zur Theorie der Lichtstrahlen, *Annalen der Physik* 255 (3rd ser.) 18: 663–695.

―――(1891a) *Gesammelte Abhandlungen von G. Kirchhoff*, Leipzig: J. A. Barth.

―――(1891b) **V**orlesungen *über Mathematische Physik, Optik*, Leipzig: B. G. Teubner.

Kline, Morris (1972) *Mathematical Thought from Ancient to Modern Times*, New York: Oxford University Press.

Kong, Jin Au (1986) *Electromagnetic Wave Theory*, New York: Wiley.

Kottler, Friedrich (1923) Zur Theorie der Beugung an schwarzen Schirmen, *Annalen der Physik* 375 (4th ser.) 71: 405–456.

―――(1965) Diffraction at a black screen. Part I: Kirchhoff's Thoery, *Progress in Optics* 4: 283–314.

Kuhn, Thomas S. (1962) *The Structure of Scientific Revolutions*, Chicago: University of Chicago Press.

Lorentz, Hendrik Antoon (1887) De l'influence du mouvement de la terre sur les phénomènes lumineux, *Archives Néerlandaises* 21: 103–176.

Lucke, Robert L. (2004) *Rayleigh-Sommerfeld diffraction vs Fresnel-Kirchhoff, Fourier propagation, and Poisson's spot*. NRL/FR/7218–04–10,101. Washington, DC, Naval Research Laboratory. Online available at http://oai.dtic.mil/oai/oai?verb=getRecord&metadataPrefix=html&identifier=ADA429355.

Maggi, Gian Antonio (1888) Sulla propagazione libera e perturbata delle onde luminose in un mezzo isotropo. *Annali di Matematica Pura ed Applicata* 16: 21–48.

―――(1914) Sul teorema di Kirchhoff traducente il principio di Huygens, *Annali di Matematica Pura ed Applicata* 22: 171-177.

Marchand, Erich W. and Emil Wolf (1964) Comparison of the Kirchhoff and the Rayleigh–1Sommerfeld Theories of Diffraction at an Aperture, *Journal of the Optical Society of America* 54: 587–594.

―――(1966) Consistent formulation of Kirchhoff's diffraction theory, *Journal of the Optical Society of America* 56: 1712–1722.

Marx, Werner (2016) Bibliometric analysis of Kirchhoff's paper. in K. Hentschel and Ningyan Zhu (eds.) Gustav Robert Kirchhoff's Treatise *On the Theory of Light Rays (1882)*, Stuttgart: Logos (Stuttgarter Beiträge zur Wissenschafts-u. Technikgeschichte, 10), here on pp. 157–167.

Michelson, Albert and F. G. Pease (1920) Measurement of the diameter of α Orionis with the interferometer, *Contributions from the Mount Wilson Observatory* 203: 249–260.

Poincaré, Henri (1889) *Leçons sur la Théorie Mathématique de la Lumière, professés pendant le premier semestre 1887–1888*, Paris: Georges Carré.

———(1892a) Sur la polarisation par diffraction, *Acta Mathematica* 16: 1–50.

———(1892b) *Théorie mathématique de la lumière II. Nouvelles études sur la Diffraction — Théorie de la dispersion de Helmholtz. Leçons professés pendant le premier semestre 1891–1892*, Paris: Gauthier-Villars.

Rayleigh, Lord (1896) *Theory of Sound*, London: Macmillan and Co.

———(1897) On the passage of waves through apertures in plane screens, and allied problems, *Philosophical Magazine* 43: 259–272.

———(1913) On the passage of waves through fine slits in thin opaque screens, *Proceedings of the Royal Society of London* 89: 194–219.

Riemann, Bernhard (1857). *Theorie der Abel'schen Functionen,* Berlin: Georg Reimer.

Rubinowicz, Wojciech Adalbert (1917). Die Beugungswelle in der Kirchhoffschen Theorie der Beugungserscheinungen, *Annalen der Physik* (4th ser.) 53: 257–278.

Saatsi, Juha and P. Vickers (2011) Miraculous Success? Inconsistency and Untruth in Kirchhoff's Diffraction Theory, *British Journal for the Philosophy of Science* 62(1) (2011): 29–46.

Schweber, Silvan Sam (1994) *QED and the Men Who Made It: Dyson, Feynman, Schwinger, and Tomonaga*, Princeton: Princeton University Press.

Seth, Suman (2010) *Crafting the Quantum. Arnold Sommerfeld and the Practice of Theory, 1890–1926*, Cambridge, Mass.: MIT-Press.

Shapiro, Alan (1973) Kinematic optics: A study of the wave theory of light in the seventeenth century, *Archive for History of Exact Sciences* 11: 134–266.

Sommerfeld, Arnold (1895) Zur mathematischen Theorie der Beugungserscheinungen, *Nachrichten von der Königl. Gesellschaft der Wissenschaften zu Göttingen* 1: 338–342.

———(1896) Mathematische Theorie der Diffraction, *Mathematische Annalen* 47: 317–374.

———(1954) *Optics. Lectures on Theoretical Physics, Vol. IV.* New York: Academic Press.

———Ed. (2004) *Mathematical Theory of Diffraction*, Basel: Birkhäuser.

Stokes, George Gabriel (1845a) On the Aberration of Light, in G. Stokes: *Mathematical and Physical Papers,* Cambridge: Cambridge University Press, vol. 3: 134–140.

———(1845b) On the Constitution of Luminiferous Ether, Viewed with reference to the Phenomenon of the Aberration of Light, in G. Stokes: *Mathematical and Physical Papers.* Cambridge: Cambridge University Press, vol. 3: 153–156.

———(1845c) On the theories of the internal friction of fluid in motion, and of the equilibrium and motion of elastic solids, in G. Stokes: *Mathematical and Physical Papers,* Cambridge: Cambridge University Press, vol. 2: 243–328.

———(1856) On the dynamical theory of diffraction, *Transactions of the Cambridge Philosophical Society* 9: 1–62.

———(1883) *Mathematical and Physical Papers.* Cambridge: Cambridge University Press.

Prof. Jed Z. Buchwald, Ph.D.
California Institute of Technology.
1200 East California Boulevard,
Pasadena, CA 91125, USA
Email: buchwald@caltech.edu

Assoc. Prof. Chen Pang Yeang, Ph.D.
Institute for the History and Philosophy of
Science and Technology, Univ. of Toronto
91 Charles Street West, Victoria College,
Room 316, Toronto, Ontario M5S 1K7
Canada
chenpang.yeang@utoronto.ca

Why Kirchhoff's Approximation Works, by Peter Vickers

Abstract

For over one hundred years the question has been asked about why Kirchhoff's hypotheses concerning single slit diffraction work so well, despite (i) being mutually inconsistent, and (ii) departing significantly from what we now know to be the case. We are finally in a position to answer this question. Drawing on recent research into the role of inconsistency in science, and some recent work by Geoffrey Brooker, this paper shows why there is nothing surprising (let alone miraculous) about Kirchhoff's predictive success. Inconsistent and (very) false assumptions can, in certain circumstances, lead to impressive, quantitatively accurate predictions. Kirchhoff's theory is just one striking example of this phenomenon.

Introduction

It is usual to expect false assumptions to lead to false predictions. Or, at the very least, one would expect false assumptions to lead to predictions which are some distance from 'perfect.' Consider, for example, the fact that we predicted a solar eclipse on 20th March 2015, beginning at 09.09 and 32 seconds over the Atlantic, and ending at 10.21 and 20 seconds, near the North Pole. In fact, we even knew in advance the exact positions of the start and end points. How did we know this with such accuracy, down to the second? There is a simple story to tell here: we knew all of the relevant information concerning how things were at some previous time with extreme accuracy (e.g. the relative positions of the Sun, Moon, and Earth). And we knew exactly how the relevant variables were going to vary over time. To put it another way, we started with true assumptions, and we brought these assumptions together to derive a prediction. Since the starting assumptions were true, and we used truth-preserving inferences, we knew that the prediction would be true. What would have happened if one or more of our starting assumptions was false? Well, even if it had been just a little bit false — e.g. we had the position of the Moon slightly wrong — then that would have changed the prediction significantly. In other words the prediction would have been significantly wrong, and there would have been many disgruntled tourists on the Faroe Islands on 20th March 2015.

Thinking about cases like this one, it is easy to imagine that these lessons generalize: scientific theories, in order to be successful, *must* be true (or at least very close to truth). Thus we find some philosophers favoring a 'no miracles' argument for the truth of our successful scientific theories. This argument can be summarized as follows:

1. Our best scientific theories are extremely successful.
2. It would be a miracle if our theories were extremely successful, but weren't even approximately true.
3. There are no miracles.
4. Therefore, our best scientific theories must be true, or at least approximately true.

Much has been said of this argument,[1] but the basic idea remains appealing to many philosophers of science (and scientists too). However, we do find many examples in the history of science which don't seem to fit into this picture. Thus many *historians* of science — those who know the history of science intimately — reject the 'no miracles argument.' Consider as one example Sommerfeld's prediction of the fine structure of the spectral lines of hydrogen.[2] Sommerfeld's predictions were certainly quantitatively impressive (even by modern standards), but his route to that prediction made heavy use of hypotheses concerning relativistic effects on electron

---

[1]A reader unfamiliar with this literature is directed to A. Chakravartty: Scientific realism, *The Stanford Encyclopedia of Philosophy*, spring (2014), Edward N. Zalta (ed.). URL = <http://plato.stanford.edu/archives/spr2014/entries/scientific-realism/>. One useful article which is *not* in Chakravartty's bibliography is S. Fitzpatrick: Doing Away with the No Miracles Argument, forthcoming in D. Dieks and V. Karakostas (eds.), *Recent Progress in Philosophy of Science: Perspectives and Foundational Problems*, Springer. Cf. J. Saatsi & P. Vickers: Miraculous Success? Inconsistency and Untruth in Kirchhoff's Diffraction Theory, *British Journal for the Philosophy of Science* **62**, no. 1 (2011): 29–46.

[2]See P. Vickers: 'Historical Magic in Old Quantum Theory?' *European Journal for Philosophy of Science* **2**, no. 1 (2012): 1–19.

trajectories that have no place whatsoever in the modern theory. Thinking about this case, Helge Kragh writes as follows:

> By some sort of historical magic, Sommerfeld managed in 1916 to get the correct formula from what turned out to be an utterly inadequate model ... [This] illustrates the well-known fact that incorrect physical theories may well lead to correct formulae and predictions.[3]

The words "historical magic" leave the reasons for the success a mystery, but Kragh is clear on one thing: this sort of thing really *does* happen in science. Successful predictions from (very) false starting assumptions are a scientific reality. If you want to call this a 'miracle', then it turns out there *are* miracles; premise 3, above, is false.

Kirchhoff's theory of single slit diffraction seems like just such another case where we get excellent scientific success from "an utterly inadequate model". In this case we find again that Kirchhoff's theory is 'far from the truth' (whatever that phrase could reasonably be taken to mean), but he gets near-perfect predictions anyway. More dramatically still, Kirchhoff's theory isn't even logically consistent. The inconsistency of the theory can be easily identified. Kirchhoff makes use of a certain assumption concerning the behavior of the light within the aperture, then derives a conclusion which tells us the light is doing something significantly different within the aperture to what that very assumption states. To put it simply, he makes use of assumptions A, B, C, ... and reaches a conclusion ¬A. As Heurtley put it in 1973:

> A problem of continuing interest in scalar diffraction theory is why the mathematically inconsistent theory of Kirchhoff predicts results that are in substantial agreement with experiment.[4]

So here we have something of a mystery: to explain how such success is possible, without any need for 'miracles'. In the next section I turn to the inconsistency problem; in the penultimate section I turn to the problem of getting true predictions from (merely) false assumptions, and then finally we reach a concluding discussion.

---

[3] Helge Kragh: The fine structure of hydrogen and the gross structure of the physics community, 1916–1926, *Historical Studies in the Physical Sciences* **15** (1985): 67–125, p. 84.

[4] J.C. Heurtley: 'Scalar Rayleigh-Sommerfeld and Kirchhoff diffraction integrals: A comparison of exact evaluations for axial points,' *Journal of the Optical Society of America* 63, 8 (1973): 1003–1008, quote from p. 1003.

Successful, Inconsistent Science

If one starts from an inconsistent starting point, then it is possible, using deductive logic, to derive absolutely any proposition. This is known as ECQ (*Ex Contradictione Quodlibet*). The derivation goes as follows. Starting with the inconsistent assumptions, derive a contradiction. From this contradiction A&¬A infer both conjuncts: A and ¬A. Now from A we may infer A ∨ B for any arbitrary B. (This is a simple rule of deductive logic, allowed on the basis that if A is true then A ∨ B must also be true.) Now from ¬A and A ∨ B infer B, and we have reached our arbitrary conclusion. But when it comes to inconsistency one doesn't need ECQ to get the mystery off the ground. By common sense, if somebody tries to persuade us of anything starting with inconsistent assumptions we will have no faith in any conclusions that person reaches. Thus one would expect inconsistent science to be useless science. As Karl Popper put it in 1959: "[Consistency] can be regarded as the first of the requirements to be satisfied by *every* theoretical system, be it empirical or non-empirical."[5]

Of course, in certain circumstances, one might start with an inconsistent set of assumptions, but then forget about some of them and reason only with the others. Under *these* circumstances one might end up reasoning with a set of consistent, and in fact true assumptions, even though in a sense one *started* with an inconsistent set. But this isn't what happened in the Kirchhoff case: Kirchhoff clearly *makes use of* a set of inconsistent assumptions when he derives his diffraction formula.[6]

But even when one makes use of *all* of a set of inconsistent assumptions in the same piece of reasoning, there are circumstances in which success is to be expected. Peter Smith considers the theory Q_G, which consists of some extremely basic truisms of mathematics and Goldbach's conjecture, that any even number is the sum of two primes.[7] If Goldbach's conjecture is

---

[5]K. Popper: *The Logic of Scientific Discovery*. London: Hutchinson and Co., 1959, §24.

[6]See e.g. Max Born & Emil Wolf: *Principles of Optics*, 7th (exp.) ed., Cambridge: Cambridge University Univ. Press, 1999, chap. 8; or see P. Vickers: *Understanding Inconsistent Science*. Oxford: Oxford Univ. Press, 2013, sec. 7.5 for my own reconstruction of Kirchhoff's derivation. Of course, the full story of Kirchhoff's derivation is more complicated (see the other articles in this volume). But these simplified reconstructions suffice for the points to be made here.

[7]P. Smith: *An Introduction to Gödel's Theorems*. Cambridge: Cambridge University Press, 2007.

false, then this is an inconsistent set. Suppose that this is the case, because there is *one* even number in the set of all even numbers that is *not* the sum of two primes. In that case, we have an inconsistent set of assumptions, but for nearly *all* possible derivations we could make using these assumptions we will reach true conclusions. To reach any arbitrary conclusion via ECQ we would have to first derive a contradiction — but that is something nobody knows how to do (even if it *is* possible).

Turning away from this toy case, there is a general lesson here: scientists are not deductive machines. As Feyerabend put it back in 1978: "The objection that a contradiction entails every statement applies to special systems of logic, not to science which handles contradictions in a less simpleminded fashion."[8] Compare also Paul Weingartner from 1994: "It is an important observation that scientists, if they speak of consequences of scientific theories, do have in mind something much more restricted than that what logic permits to be an element of the consequence class."[9] For any scientist, or anyone who has seriously studied some science, this is an obvious point. And the reason consequences are 'restricted' is simple enough. Within scientific reasoning, in virtually all circumstances, one knows from the outset that not *all* of one's assumptions are candidates for 'truth.' That is to say, it is virtually always the case that some approximations or idealization assumptions have been used somewhere along the way. And in these circumstances one should not take seriously the results of all possible truth-preserving inferences. Truth-preserving inferences lead us to truth *if* one starts with truth; it is not clear where they take us if one *doesn't* start with absolute truth.

We are here moving away from the question of how inconsistent science can be successful, and towards the broader question of how *any* false science can be successful. But there is a special problem with inconsistency which deserves discussion before we move on. It has sometimes been said that it is no surprise that one can reach true conclusions when one has an inconsistent starting point: after all, one can reach *any* conclusion whatsoever if one has an inconsistent starting point, via ECQ (as explained above). But whilst it is true that one *can* reach any conclusion,

[8]P. Feyerabend: 'In Defence of Aristotle,' in G. Radnitsky & G. Anderson (eds.), *Progress and Rationality in Science*, Dordrecht: Reidel, 1978, p. 154.
[9]P. Weingartner: 'Can There Be Reasons for Putting Limitations on Classical Logic?' in P. Humphreys (ed.) *Patrick Suppes: Scientific Philosopher*, vol. 3, Dordrecht: Kluwer, 1994: 89–124, quote from p. 95.

it is something that would never happen in practice. To reach any arbitrary conclusion one wishes from an inconsistent starting point, one *must* first reach explicit contradictories — that is the point of the 'C' in 'ECQ.' And then, having reached the contradictories, one would have to *continue making inferences*. To put it bluntly, one would have to be off one's rocker to suggest that this could ever happen in real scientific practice. Rather, in real scientific practice, whenever one reaches a contradiction one knows that something has gone wrong, and one looks back to see what might be to blame. Even if one originally thought one had started with true assumptions, upon deriving a contradiction one knows that at least one assumption must be false (assuming sound reasoning). And so one knows immediately that truth-preserving inferences cannot be trusted to lead us to truth.

This all leads to the conclusion that there is no *special* problem with inconsistent science. Instead the inconsistency problem is just a special case of a broader problem: how can successful reasoning proceed when one starts with a set of assumptions at least some of which are false (including the case where the set is inconsistent)? Under what circumstances *do* we take seriously the results of truth-preserving inferences when one has not started with truth? And in particular, how can one get extremely accurate predictions when one has not started from truth?

Success from Falsity

It is *very* common in physics to make use of assumptions one knows to be false. In such circumstances, one is simply careful about *any* inferences one makes. If one starts from assumptions at least some of which are false, then it will definitely be possible to derive absurd conclusions if one goes out of one's way to do so. But this is obviously not how science works! Let us consider the sense in which Kirchhoff's assumptions were *not* true, and ask how it was nevertheless possible for him to derive impressive, accurate predictions.

Consider Figure 15. The most important assumptions for present purposes are Kirchhoff's assumptions concerning the behavior of the light across the aperture (labeled A), and immediately behind the screen (labeled B). Here Kirchhoff made some crucial approximation assumptions. He assumed that the light behavior across aperture A would be exactly as if the screen were not present. And he assumed that the light behavior

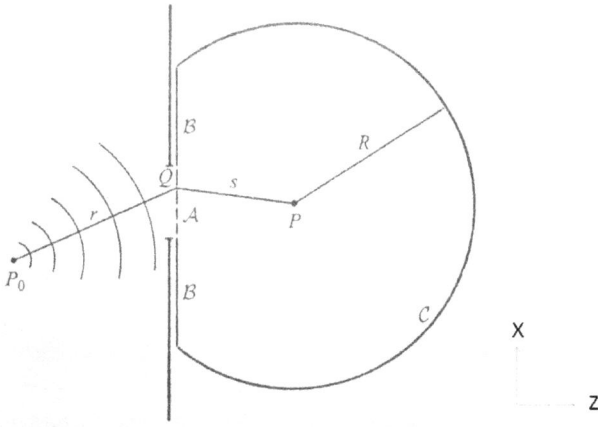

Figure 15: A birds-eye view of the theoretical situation. P_0 is the source of the light, and P is the point beyond the screen at which we want to know the light intensity. In addition Q is a point in the aperture whose contribution we are considering at a given time, r is the distance from P_0 to Q, and s is the distance from Q to P. An imaginary surface of integration S is comprised of A (the aperture), B (part of the screen), and C (part of a sphere of radius R which has P at its centre). The figure suggests non-normal incident radiation, but I will here consider only normal incident radiation. (Figure taken from Born & Wolf 1999, p. 421.)

immediately behind the screen (the sections labeled B in the figure) would be exactly as if the screen did not have an aperture. Or, to put it another way, he assumed that the radiation across the aperture would simply be a piece cut from the incident wave, and the radiation and its $+z$ derivative behind the screen would both be zero. (Coordinates x and z are given in Figure 15; the y direction is into/out of the page.) Now of course Kirchhoff knew that these assumptions would be false of any real-life experimental situation. He knew, in particular, that any real screen would have a finite thickness which would affect the light behavior at both A and B. In these circumstances, how did he expect to derive any meaningful/accurate results?[10]

The answer is obvious: he thought that if a screen were used which was *very* thin (and very opaque) then his assumptions would be *approximately*

---

[10]To be more precise, Kirchhoff doesn't specify the thickness of the screen. But his reasoning indicates that he assumed the thickness to be negligible, even if his diagram taken on its own might suggest a non-negligible thickness.

true. And so he expected that, starting with approximately true (and also true) assumptions, and reasoning deductively, at least some of his predictions (possibly most of his predictions) would be approximately true. If the screen used in the experiment were thin *enough* then, just possibly, the effects of the width of the screen would be negligible, and the predictions would be extremely accurate.

We all know what happened in practice: Kirchhoff's predictions were, for the most part, *extremely* accurate.[11] Where do the predictions break down? One place we know that they break down is very near to the aperture. Most dramatically we know that they give results *at* the aperture A in conflict with Kirchhoff's very assumptions — as already noted in the previous section. But Kirchhoff would have brushed this problem aside as a mere artefact of the approximation assumptions he had made. In general, when scientists know that they have started from only *approximate* truth they do not blindly trust their inferences, but instead examine what is inferred, checking it empirically and conceptually. Empirically and conceptually Kirchhoff's predictions near to the aperture carry zero scientific value, and should be ignored. By contrast most other predictions are accurate enough to be extremely valuable from the perspective of both science and technology.

This might have been the end of the story but, as is well known, there is a startling twist to the tale. Kirchhoff's approximation hypotheses are absolutely *not* approximately true in the way Kirchhoff assumed.[12] One way to see this is to use Maxwell's equations to get a (very) accurate story concerning the behavior of the radiation across the aperture when the screen has negligible thickness. Consider E_y — the component of electric field **E** in the y direction — and the difference between this amplitude at the aperture and the incident E_y-amplitude.

Kirchhoff assumed these two things were the same, and so we get the flat horizontal line in Figure 16 (the ratio of one to the other is 1). But the Maxwell equations tell us that, in truth, we should have a wavy line in Figure 16. The E_y-amplitude behavior at the aperture is *not* simply a

---

[11]See G. Brooker: Diffraction at a single ideally conducting slit, *Journal of Modern Optics* 55, 3 (2008): 423–445, for several dramatic figures showing just how accurate the predictions are. For most angles of diffraction, and sufficiently far from the aperture, they are essentially 'perfect'.

[12]At least, if we consider natural intuitions concerning what 'approximately true' could reasonably mean.

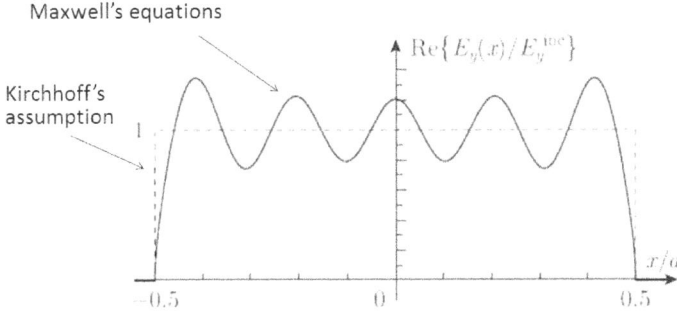

Figure 16: Comparison of Kirchhoff's assumption of a "flat" amplitude function across an aperture of width a with the amplitude function derived from Maxwell's equations. (Adapted from G. Brooker: *Modern Classical Optics.* Oxford: Oxford University Univ. Press, 2003, p. 71.) 'Re' refers to the 'real part,' of course; there is also an imaginary part not considered here.

'piece cut from the incident wave.' And this is not a good approximation, either.

We now have a new puzzle: how could Kirchhoff reach such perfect predictions when he was so wrong about the light behavior at the aperture and behind the screen? In particular, why didn't Kirchhoff's mistake concerning the light behavior within the aperture translate into a mistake concerning his predictions (predictions which were reached making direct and explicit use of the false assumptions)?

Geoffrey Brooker has done more than most to answer this question. In his book *Modern Classical Optics* he states that in this case "nature has been unusually kind to us", the thought being that it is not 'usual' that one would get accurate predictions from such a misguided starting point. But although it is unusual there is no 'miracle', and we can understand what has happened.[13] For simplicity, consider the case where the radiation incident on the slit is polarized such that the **E**-field is in the $(+/-)y$ direction, and the **B**-field is in the $(+/-)x$ direction, with direction of travel in the $+z$ direction (see Figure 17).

For this polarisation case Brooker shows that, within the aperture, in addition to the incident radiation, we also have two cylindrical waves travelling from the upper and lower jaws of the slit in the $(+/-)x$

---

[13] It is interesting to note that here we have a physicist asking, and attempting to answer, the same questions as philosophers of science (even if the style of expression is sometimes quite different).

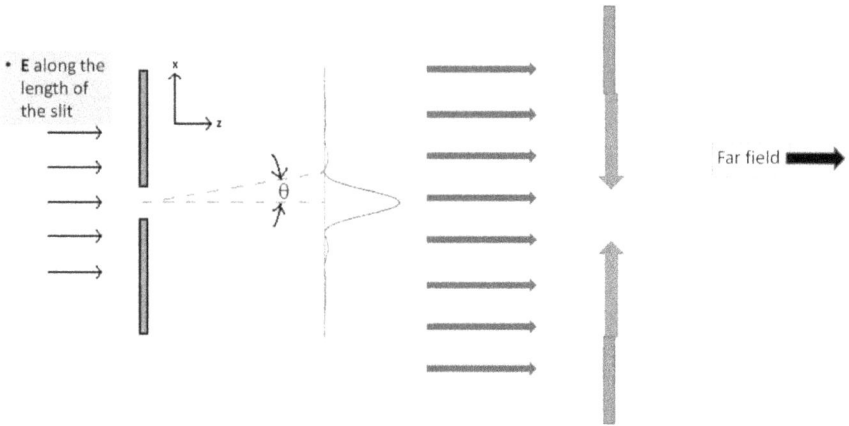

Figure 17:   Left hand side: Incident light approaches the aperture from the left, polarized such that **E** is in the $(+/-)y$ direction, and **B** is in the $(+/-)x$ direction. A diffraction pattern appears at large $+z$, in the 'far field'. Right hand side: Looking more closely at the aperture, it turns out that the behavior of the light there can be separated into three parts: (i) a piece cut from the incident wave, travelling in the $+z$ direction, (ii) a wave (vertical arrow) travelling in the $+x$ direction from the lower jaw, (iii) a wave (vertical arrow) travelling in the $-x$ direction from the upper jaw.

direction.[14] Since (within the aperture) they are travelling in the x direction there are **E** and **B** amplitudes in the y and z directions, and in particular we find that there are **E** amplitudes in the y direction which are not negligible. This explains why E_y differs from the Kirchhoff assumption in Figure 16, above. What the solid line in Figure 16 really gives us is an interference between three different waves: the incident wave, and the two waves travelling from the jaws in the $(+/-)x$ direction.

Now of course, we know that Kirchhoff had no idea about the two cylindrical waves travelling in the $(+/-)x$ direction. But there is an obvious sense in which it didn't matter that Kirchhoff missed this phenomenon. It has long been known that waves on a collision course pass through each other, emerging after interaction in exactly the same form as if the other waves had never existed. Thus the incident radiation at the aperture, despite being (in a sense) 'interfered with' at the aperture by the cylindrical

---

[14]Brooker [2008], sec. 5. The sense in which they are cylindrical waves is simply that they originate in a 'line source,' namely, each jaw of the aperture when we consider the three-dimensional version of events, with the aperture and screen extending into/out of the page.

waves, emerges from the aperture and continues to the far field exactly as if the cylindrical waves had never existed. In other words, if what we are interested in is the diffraction behavior *in the far field*, then we can ignore the interference effects of the cylindrical waves at the aperture. Kirchhoff didn't *ignore* this, of course — he just didn't know about it. But the end result is the same.

We might wonder whether it is really the case that the non-incident radiation in the aperture is moving *only* in the x direction. Might some of it be moving more or less in the $+z$ direction? We can see that, in fact, within the aperture, it is moving directly in the x direction. If it *were* moving directly in the x direction we would expect no **E** or **B** component in the x direction. And that is indeed what we do find. In Figure 18 we see the B_x behavior within the aperture and behind the screen (for the polarisation case under discussion, with incident **E** in the y direction). The B_x amplitudes within the aperture are precisely the incident B_x amplitudes, exactly as Kirchhoff expected. This is not because the incident radiation is the only radiation present, of course. It is just that the non-incident radiation (from the cylindrical waves) is travelling in the $(+/-)x$ direction, and so can't possibly contribute anything to the B_x amplitudes (*mutatis mutandis* for E_x).

At this point we have a solid explanation of why Kirchhoff's assumptions about the behavior in the aperture didn't impact upon his far field diffraction predictions. But it should be obvious that the radiation coming

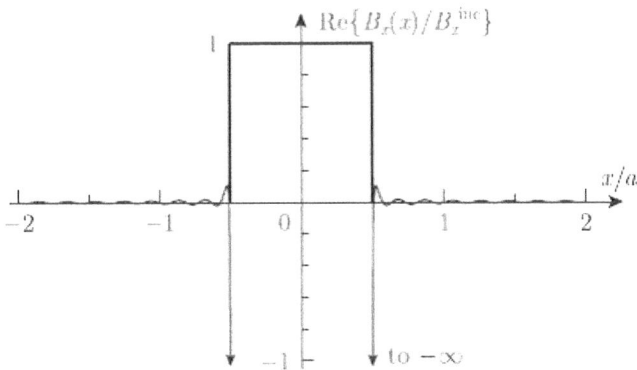

Figure 18: Brooker [2008], fig. 8a, gives the x component of **B** within the aperture (for the polarisation case where the incident radiation has **B** in the x direction). Kirchhoff's assumption is almost exactly right here: the B_x behavior in the aperture is a piece cut from the incident wave.

from the jaws is not directed *purely* in the x direction. Instead, some of it heads out into the $+z$ half-plane at angles of θ such that $0 < \theta < \pi$.[15] How is it the case that Kirchhoff's neglect of *this* phenomenon didn't impact upon the accuracy of his predictions?

The short answer to this is that the line-source radiation emanating from the jaws reduces in intensity as one considers values of θ increasingly near to $\pi/2$ (towards the far field), and also diminishes with distance relatively rapidly in that direction (relative to the behavior of the incident radiation). Ideally we would like to separate two equations: one for the incident radiation and its behavior beyond the aperture, and another equation for the jaw-source radiation and its behavior beyond the aperture. Because if we had an equation for the behavior of the cylindrical wave in the far-field direction, we could simply look and see whether this radiation would be negligible or not in that direction.[16]

Several different attempts have been made in this direction. E.g., Born and Wolf consider diffraction by a half plane, and attempt to separate the behavior into a contribution from the "geometrical optics field" and a line-source contribution coming from the jaw. They call the latter the "diffraction field" and write that this is "simply the field which must be added ... to give the complete field."[17] The equation they provide for this diffraction field (equation (34)) is:[18]

$$E_z^{(d)} \sim \sqrt{\frac{2}{\pi}} e^{\frac{1}{4} i \pi} \frac{\sin \frac{1}{2}\alpha_o \sin \frac{1}{2}\theta}{(\cos \theta + \cos \alpha_o)} \frac{e^{ikr}}{\sqrt{kr}} \qquad (34)$$

Unfortunately this equation can't help us here. It is highly approximated, and only valid under certain conditions. Just one example of this restricted validity is: "When $\cos\theta + \cos\alpha_o$ approaches zero, the approximation (34) breaks down."[19] But that is just the situation we are

---

[15]Angles are here measured in radians, in line with Born and Wolf [1999]. As we look at Figure 15, $\theta = 0$ is taken to be straight up, and $\theta = \pi$ is straight down. $\theta = \pi/2$ is in the $+z$ direction, directly towards the far field.

[16]The term 'cylindrical wave' could be misleading here: the cylinders are certainly not 'perfect cylinders.' And how much they 'look like' cylinders might well depend on the distance from the line source. In addition they might 'look like' cylinders for one field component, but not for another.

[17]Born and Wolf [1999] chap. 11, p. 651.

[18]It is crucial to note how the authors choose to label the axes. Their 'z' is Brooker's 'y,' and vice versa. In particular, when they write E_z, that corresponds to what Brooker 2003, 2008 writes as E_y. In this paper I follow Brooker's labeling conventions, but I present equation (34) using Born & Wolf's conventions.

[19]ibid.

interested in, because we are interested in cases where we have normal incident radiation ($\alpha_o = \pi/2$)[20] and also where we are looking towards the far-field ($\theta \approx \pi/2$). In addition, the Born and Wolf discussion is not exactly relevant for present purposes: they use a geometrical optics 'ray' approach to diffraction as opposed to Kirchhoff's Huygens secondary source approach.

Another attempt to separate the contributions to diffraction comes from Pyotr Yakovlevich Ufimtsev, who discusses the case of diffraction of a plane wave by a perfectly conducting wedge (with a half-plane as a special case).[21] In this case an exact treatment is separated into the Kirchhoff contribution and a correction term which describes what might be called a "fringe wave" (see Section 4.1 of Ufimtsev 2014):

$$f^{(1)}(\varphi, \varphi_o, \alpha) \frac{e^{i(kr+\pi/4)}}{\sqrt{2\pi kr}},$$

where $f^{(1)}(\varphi, \varphi_o, \alpha)$ is given by trigonometric functions and is finite everywhere. And this correction term has negligible effect at the boundary of incidence (for example), thus helping to explain why Kirchhoff's formula is so accurate in that direction.

At the same time, this is certainly not a rigorous account of the contribution from the jaw, and indeed such an account may not even be possible. But such an account may not really be necessary, anyway. One can't reasonably doubt that the contributions from the jaws have negligible effect on the far-field. It is obviously the fact that this radiation *is* negligible in the far-field, which accounts for the fact that Kirchhoff's predictions were 'perfect' despite him not taking it into account.

I have here focused on Kirchhoff's assumptions concerning the light behavior at the aperture. There is also the question of Kirchhoff's assumptions concerning the light behavior immediately behind the screen. The story is very similar here. Consider the other polarisation case, where **B** is in the y direction and **E** is in the x direction. We again get cylindrical waves travelling along the x axis, but now they travel *away* from the aperture instead of into it (see Figure 19).

Thus we *don't* get a 'perfect shadow' behind the screen as Kirchhoff assumed. But again, these disturbances are not travelling towards the far

---

[20] See, e.g., Born & Wolf 1999, p. 649, fig. 11.9.
[21] P.Y. Ufimtsev: *Fundamentals of the Physical Theory of Diffraction*, 2nd edition. John Wiley & Sons, 2014. Many thanks to Ning Yan Zhu for bringing this to my attention.

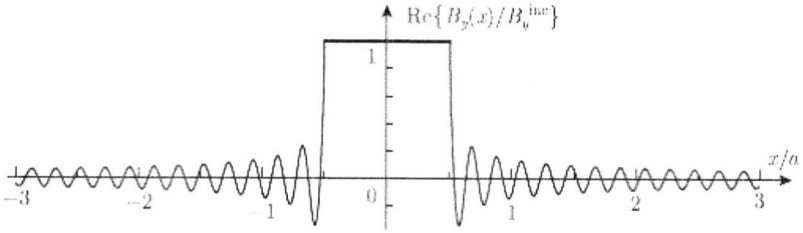

Figure 19:   Brooker [2008], fig. 3a gives the y component of **B** within the aperture and immediately behind the screen (for the polarisation case where the incident radiation has **B** in the y direction). Kirchhoff's assumption is exactly right within the aperture, but wrong behind the screen.

field, and thus have a completely negligible effect on the far-field diffraction pattern. Some disturbances do travel a little more towards the far-field, but the significance of these disturbances diminishes the nearer we get to $\theta = \pi/2$. Thus Kirchhoff gets away with the assumption that there is no light behavior immediately behind the screen: nothing immediately behind the screen is relevant vis-á-vis the far-field diffraction pattern.

Discussion

We have answered Heurtley's question, introduced in the first section above. Simply put, it needn't be surprising that "the mathematically inconsistent theory of Kirchhoff predicts results that are in substantial agreement with experiment." An inconsistent set of assumptions can consist of a large set of true assumptions and just one other assumption which is itself approximately true. In these circumstances if one uses truth-preserving inferences one would expect many true conclusions, and also many approximately true conclusions. One would also expect some conclusions which are 'way off.' And we do indeed find all of these things with Kirchhoff's theory.

But there is a harder question, the question of how Kirchhoff could get results that are in substantial agreement with experiment when some of his hypotheses are *not* approximately true. Especially when we consider the fact that Kirchhoff made explicit use of his 'way off' assumptions to reach his impressive predictive successes. The answer provided here comes from the fact that the true state of affairs can be separated into aspects which have an impact on the far-field diffraction pattern and aspects which do not, combined with the fact that Kirchhoff was ignorant only of aspects which do *not* matter to the far-field effects. To be more specific, Kirchhoff overlooked the contribution of the cylindrical waves coming from the aperture jaws to the aperture field amplitudes (and the amplitudes immediately behind the screen). But the fact that these waves make a significant difference to the screen-plane amplitudes does not imply that they also make a significant difference to the far-field. The jaw-source radiation which affects the screen-plane amplitudes is radiation travelling in the $(+/-)x$ direction, and it can hardly turn a corner and travel towards the far-field.

Suppose we say, then, that the screen-plane field amplitudes are due to three waves: the incident wave and the radiation coming from the two jaws. We might then be tempted to say that Kirchhoff was *not* wrong about the incident radiation behavior at the screen plane. Rather, he just overlooked some *other* behavior that also happens. One might be tempted to say that, within the aperture, the incident wave *does* exist in precisely the form Kirchhoff supposed, it's just that some other waves exist there as well, in exactly the same place. And these waves are just passing through each other at that point like ghosts in the night.

Certainly people sometimes talk about waves 'passing through' each other. But in addition people often talk about waves 'interfering'. We might be tempted then to ask "So which is it? Do they pass through, or do they interfere"? One gets the right answer if one thinks about two waves meeting at the centre of a piece of rope held taut at each end. Clearly it isn't possible for the vibrations in the rope to *truly* pass through each other without any sort of interference. And, indeed, if one thinks about the fact that the amplitude of the rope increases in size at the point of interference, it is clear we do have genuine interference (if the waves *truly* passed through each other, we would expect no such increase in amplitude). So the talk of 'passing through' is colloquial shorthand: it refers to the fact that, *following* the interference at the meeting point, the two waves emerge on the other side in exactly the same form as before, *as if* the other wave had never existed. No doubt this is rather odd behavior, and the word 'interfere' still does not sit comfortably. In just about any other context one can imagine, if one thing interferes with another, that means that the first thing emerges after the interference somehow *different* to how it would have been if the interference had never taken place. But this is just a feature of waves that one has to get used to, and the word 'interference' in this context is here to stay.

The upshot of this for the Kirchhoff discussion is that we should *not*, strictly speaking talk as if there are three waves 'present' within the aperture. At each point within the aperture there is only one amplitude, not three existing in the same place.[22] Accordingly, Kirchhoff *was wrong* in his assumption concerning the aperture amplitudes (and also the amplitudes behind the screen). But when waves meet it happens that the interference that takes place also undoes itself perfectly. And this is ultimately the reason why Kirchhoff could be so successful in his predictions despite getting the aperture amplitudes so wrong.

Turning back to the scientific realism debate, many contemporary realists put a lot of emphasis on a strategy known as *selective realism*, where one identifies 'idle' parts of scientific theories vis-á-vis some prediction, and 'working' parts.[23] The idle parts are the parts of a theory which don't feature in generating the prediction, and the working parts are the parts which do. This is thought to be a natural way to explain how, sometimes, a theory which contains significant falsehoods can nevertheless be successful. Such realists sometimes even hope to identify idle parts of contemporary scientific theories, so that we might consider withdrawing any doxastic commitment to these parts.

At first it looks like this manoeuvre can be applied to the Kirchhoff case, identifying the incident wave as 'working', and the two cylindrical waves as 'idle' vis-á-vis the far-field diffraction pattern. But in fact things are not so straight forward, and the realist may need to find another way to accommodate this case. If we look at Kirchhoff's derivation, he assumes certain amplitudes within the aperture, and these amplitudes seem quite obviously to be (i) 'working' (they are 'doing work' in his derivation), and (ii) significantly false. It doesn't seem feasible — to this author at least — to claim that Kirchhoff's amplitude assumptions contain some working parts and some idle parts. Of course, Kirchhoff's amplitude assumption can be reconstructed mathematically as 'made up of' different contributions. And *in hindsight* we can think of the amplitudes as 'made up of' an incident wave and two cylindrical waves coming from the jaws. But if we ask "Which assumptions did Kirchhoff make which are idle"? we find ourselves at a loss. The point of the selective realist strategy is to separate the parts of a

---

[22]Although of course there is the **E** amplitude and the **B** amplitude, and we can talk about different components of these amplitudes.

[23]There are various different variations on this manoeuvre, but the differences don't matter for present purposes. See Chakravartty [2014], §2.3) for an entry to this literature.

theory we ought to believe from the parts that we ought *not* to believe.[24] Given the way Kirchhoff's amplitude assumptions featured in his derivation concerning light intensities in the far-field, it seems that, according to selective realism, one *should* commit doxastically to those amplitudes. And yet with hindsight we know that one would then believe a falsehood. Things would be different if the incident radiation amplitudes truly 'existed' in the aperture (in addition to the cylindrical wave amplitudes). But, as already discussed, this is the wrong way to conceive of the radiation activity within the aperture. One can speak in that way — and people often do — but such talk is only a convenient shorthand or *façon de parler*, and cannot be taken literally.

This is not to say that the realist cannot make any sense of this case. Certainly in hindsight the realist can find a nice way of explaining how Kirchhoff's theory could be successful, despite being false in certain important respects. But it does suggest that there can be exceptions to the realist's claims concerning the link between scientific success and scientific truth. Assumptions which are false — significantly so — can sometimes be employed to generate a prediction which is 'perfect'. This sort of phenomenon should encourage us to be cautious (at the very least) when we try to make the jump from scientific success to having confidence that our theories are true, or even approximately true, or even that certain parts of them are true/approximately true. So many times in the history of science scientists have (with good reason!) been extremely confident concerning their theories, and later the theories have been replaced by a successor theory which looks quite different. One only need reflect on the success of Bohr's theory of the atom, in particular Bohr's tremendous success in explaining and predicting the spectral lines of hydrogen and ionized helium. As Abraham Pais writes:

> Up to that time no one had ever produced anything like it in the realm of spectroscopy, agreement between theory and experiment to five significant figures.[25]

---

[24] An assumption might be idle in one derivation, but not in another. If it is doing work in *any* derivation leading to significant empirical success, then it deserves doxastic commitment. Or at least, that is the short story, for the majority of scientific realists.

[25] Abraham Pais: *Niels Bohr's Times*, Oxford: Oxford University Press, 1991, p. 149.

No wonder, then, that Einstein was moved to remark, "This is a tremendous result. The theory of Bohr must then be right".[26] Well, it turns out that tremendous results/accurate predictions and 'being right' are quite different things. Even the greatest scientific minds have too quickly jumped to make the tempting inference from success to truth. Kirchhoff's success is just such another cautionary tale in the history of science.

Senior Lecturer Dr. Peter Vickers, BSc, MA, PhD
Department of Philosophy
Durham University
50/51 Old Elvet
Durham DH1 3HN, United Kingdom
Email: peter.vickers@durham.ac.uk

---

[26] Pais 1991, p. 154.

A Brief Bibliometric Analysis
of Kirchhoff's Paper,
by Werner Marx

Abstract

On the basis of the *Thomson Reuters* citation indexes accessible under the *Web of Science* (WoS), I carefully analyze the frequency with which the article by Gustav Robert Kirchhoff from 1882 has been cited in the literature. I also compare these formal citations of one specific article with general references to Kirchhoff. Among all of his journal articles, the one most often cited turns out to be his 1847 *Annalen der Physik* paper on electric circuit laws. By contrast, a strong interest in his 1882 article on optical diffraction has only developed since the 1960s, with annual citation rates increasing up to 12 in the first years of the 21st century, and with a very broad range of nationalities of citing authors (cf. Table 1).

Introduction

An increasingly popular way to estimate the impact of a researcher is to count the number of times that his or her articles have been cited. The number of citations is often taken as a measure of the attention an article, a researcher or an institute has attracted. Although citation numbers reflect strengths and shortcomings and are therefore frequently used for research evaluation, they cannot easily be equated with the overall significance. First of all, the final importance of the more recent articles may not yet be clear. Secondly, the results of old articles may now be so well known that they appear in textbooks rather than being cited. The question arises as to whether the impact of early pioneers of science like G.R. Kirchhoff can be quantified by bibliometric methods usually applied to present-day scientists. Carefully establishing and interpreting the citations seems to be a reasonable way to proceed.

Table 1   Distribution of the articles citing the Kirchhoff article (all versions) by country of the authors (available from
about 1972 to 2010).

Country of Author	# Citing Articles	% Citing Articles
USA	36	39.6
Turkey	10	11.0
France	6	6.6
Germany	5	5.5
England	4	4.4
Australia	3	3.3
Italy	3	3.3
Netherlands	2	2.2
Poland	2	2.2
Scotland	2	2.2
Austria	1	1.1
Brazil	1	1.1
Canada	1	1.1
Cuba	1	1.1
Denmark	1	1.1
India	1	1.1
Israel	1	1.1
Russia	1	1.1
South Korea	1	1.1
Spain	1	1.1
Sweden	1	1.1
USSR	1	1.1

Methodology

The data presented here are based on the *Thomson Reuters* citation indexes
accessible under the *Web of Science* (WoS), in particular the *Science
Citation Index* (SCI). *The Social Sciences Citation Index* (SSCI) available
since 1956, the *Arts & Humanities Citation Index* (A&HCI) available since
1975, the *Conference Proceedings Citation Index, Science* (CPCI-S) and
the *Conference Proceedings Citation Index, Social Science & Humanities*
(CPCI-SSH), both available since 1992, have been consulted in addition [1].
Furthermore, the INSPEC database for Physics, Electronics, and Computing (accessible under the database provider *STN International*) has been
used for this study [2]. The WoS and the INSPEC bibliographic records
stretch back to 1900 as the year of publication.

The WoS has two search modes: The 'General Search' mode gives access
to the articles published since 1900 (no books, no conference proceedings

unless they appear in journals) and that are covered by the so-called WoS source journals. These journals currently total about 11,000, selected by the staff of *Thomson Reuters* as contributing significantly to the progress of science. The 'Cited Reference Search' mode gives access to all references which appeared in source journal articles. The cited references are not limited to articles published in source journals and include any other published material, in particular articles not published in source journals, books, and conference proceedings. In other words: the WoS records are limited to source journal articles published since 1900, but the articles cited therein are not restricted concerning document type or publication year.

Overall Citation-based Impact

The time curve of the total number of citations of Kirchhoff's publications up to 2010 (articles as well as books and lectures) was determined on the basis of the WoS Cited Reference Search mode. The citation impact based on the time period 1840–1900 (the publication years relevant here) is shown in Figure 20. Such seminal work is often cited by mentions of the author's name or name-based items ('informal') citations [3], also called 'eponyms') instead of full references as footnotes ('formal' citations). Therefore, separate time curves of the informal citations based on WoS and the INSPEC database ("Kirchhoff" appearing in the titles, the abstracts or the keywords) are included in the figure.

The total number of records covered by the INSPEC database may be taken as a reference for the overall growth of the physics literature: Both the informal citations of Kirchhoff and the total physics literature increased by about the same amount, doubling since around 1950.

An analysis of the ensemble of 2,240 articles formally citing the works of Kirchhoff reveals a review article published 1982 [4] which has been cited 1,640 times until 2010. It may be assumed that this review introduced the early Kirchhoff articles into the citation network of the more recent physics literature. Additional proof is obtained by the number of 15 co-citations: articles which cite a Kirchhoff article simultaneously with the frequently cited review. Altogether 85 out of 2,240 citing articles of Kirchhoff's works are not covered by the *Science Citation Index* but only by the *Social Sciences Citation Index* and *Arts & Humanities Citation Index*. This can be seen as a clear indication of substantial impact beyond the natural sciences.

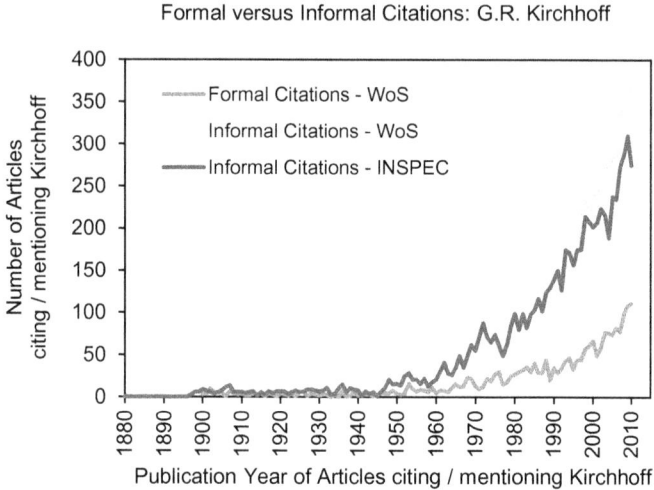

Figure 20: Time-dependent number of formal (reference-based) citations referring to the works of Kirchhoff versus informal citations ("Kirchhoff" appearing in the titles, the abstracts or the keywords) based on WoS and INSPEC. Source: *Thomson Reuters Web of Science* (WoS) and INSPEC under *STN International*.

Citations of the Article *Zur Theorie der Lichtstrahlen*

The most frequently cited Kirchhoff article appeared in the year 1847 in *Annalen der Physik* [5] (322 citations). One of the other often cited Kirchhoff publications is his article entitled *Zur Theorie der Lichtstrahlen* (citation rank 6). The impact and reception of this article is analyzed here in more depth. The article was published in 1882 in the proceedings of the Prussian Academy of Sciences, *Sitzungsberichte der Königlich Preußischen Akademie der Wissenschaften zu Berlin* [6] and in 1883 as a journal article in *Annalen der Physik* [7]. A French translation appeared 1886 in *Annales Scientifiques de l'École Normale Supérieure* [8]. According to the WoS, these three versions of the article received altogether 93 citations by 2010 (*Sitzungsberichte* 1882: 23 citations, *Annalen der Physik* 1883: 79 citations, translation 1886: 0 citations — 9 citing articles referenced both versions). Since the WoS database only starts at 1900, no routine check of earlier references to the article was possible. In PROLA, which allows citation searches in all articles of *Physical Review* from 1893 onwards, no reference to Kirchhoff's article could be traced. The only citation of Kirchhoff's article prior to 1900 known to us is by Gian Antonio Maggi [20] in 1888, actually one of Kirchhoff's students in Berlin, who had returned to Messina when he wrote his article.

The graph displaying the time-dependent evolution of a single article is sometimes called its citation history. Each article develops its own life span as an object of citation. With time, the citations per year (citation rate) normally evolve in a similar pattern: They generally do not increase substantially before one year has elapsed since publication. They reach a summit after about three years, the peak position depending somewhat on the research discipline. Subsequently, as the articles are displaced by newer ones and interest in the field wanes, their impact decreases, leading to a lower rate of citation accumulation. Finally, most articles are barely cited at all or are forgotten. Figure 21 shows the citation history after 1900 of the three versions of Kirchhoff's article taken together.

To appraise the altogether 93 articles citing the Kirchhoff article, we may take as a benchmark the citations of all articles of *Annalen der Physik* within the specific publication years around 1883: The overall number of citations of the articles published between 1880 and 1885 ranges between

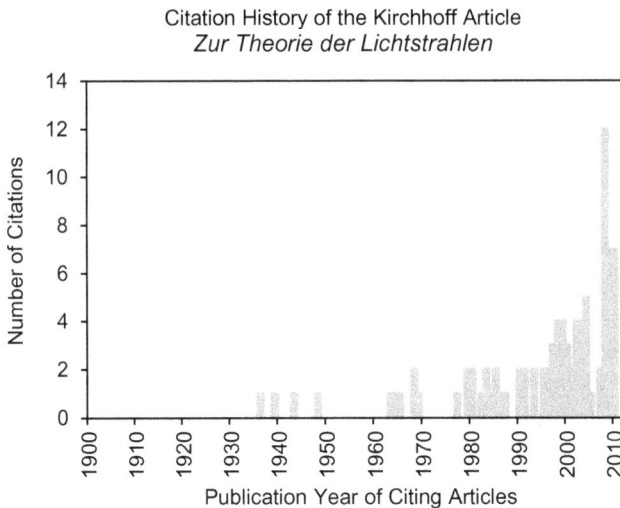

Figure 21: Time-dependent number of citations (citation history) of the Kirchhoff article *Zur Theorie der Lichtstrahlen*. Misspelled citations (incorrect with regard to the numerical data: volume, starting page, and publication year) are included here, as far as it was possible to identify them. The publication year 2008 is approximated here. Source: *Thomson Reuters Web of Science* (WoS).

700 and 1,200 per year. The Kirchhoff article is the second most-cited article published in *Annalen der Physik* in the year 1883 — the most-cited one, by L. Graetz, on thermal conductivity of fluids, is entitled *Ueber die Wärmeleitfähigkeit von Flüssigkeiten* [9]. Only two [10, 21] out of the 93 citers of the Kirchhoff article is not covered by the *Science Citation Index* but only by the *Social Sciences Citation Index* and *Arts & Humanities Citation Index*.

The first journal article to cite the Kirchhoff article past 1900 and covered by the WoS source journals did not appear before 1936 (see Figure 21). However, this citing article (as well as most of the following citing articles) did not receive enough citations to promote the Kirchhoff article significantly — with one exception: An analysis of the ensemble of 93 citing articles points to an article published in 1968 in a mathematics journal [11] which has been cited 200 times. This is a comparatively high impact in view of citation habits within the field of mathematics. That most influential citing article is a possible candidate transferrer of the Kirchhoff article into the recent literature.

In addition to the *Thomson Reuters* citation indexes, the SAO/NASA Astrophysics Data System [12] and the Physical Review Online Archive (PROLA) [13] have been consulted. Both archives reach back to before 1900. The outcome: There are 17 citers of the specific 1882/1883 Kirchhoff article accessible under SAO/NASA — 13 of these citing articles appeared since 2008. The 6 citers under PROLA could not be verified due to problems with optical character recognition (OCR) of references. The citations which possibly appeared in journals not covered (or not fully) by the WoS and within textbooks or any other material published by December 2010 were not accessible via databases and could hardly be estimated.

In contrast to the standard "canonical" time pattern mentioned above, the citations of the most often cited Kirchhoff articles are considerably delayed. Such articles have been called "sleeping beauties" [14]. The question arises as to why such early articles are cited at present. A look at the more recently published citing articles reveals that the Kirchhoff article is mostly mentioned in the historical overview given in introductions. For example: "For the solution of the wave equations concerning spatial propagation of light rays, the methods given by Kirchhoff [6] are of essential value." [15]

The evolution of citations over time is a result of two competing phenomena: the aging of the articles (obsolescence, replacement, oblivion) and the growth of the body of the scientific literature [16]. The articles covered by the SCI as well as by INSPEC increased by about a factor of one hundred throughout the 20th century. The proliferation of science implies a proliferation of potentially citable articles, resulting in increasing ratios of references per article (reference count) and therefore of the average number of citations per article (citation rate). We may speculate about how much more a citation around 1900 is worth compared to a present-day citation and may decide that citation numbers have inflated by a factor ranging between ten and a hundred. It should be pointed out that citation counts are being compared here across a time period of radically changing habits and cultures of publication and citation.

An analysis of the distribution of the citing articles by country of the authors since 1972 reveals an unexpectedly large reception by US authors. They total almost 40% of the citing authors. This is particularly striking in comparison to the Germans (see Table 1). It is valid even if the relative output of publications in chemistry and physics of the two countries is taken into account. This is the more astonishing as Kirchhoff was a famous German scientist and his publications appeared in German, which were presumably less easily accessible to many US scientists.

Furthermore, the citing articles were analyzed with regard to the research fields assigned to the journals in which they appeared. The outcome: The articles cover a vast range of subject categories with *optics* at the top, followed by the various categories of *engineering* and *mathematics*.

Obliteration by Incorporation

The works of Gustav Kirchhoff are a typical example of "obliteration by incorporation," a phenomenon first described in 1949 by the sociologist Robert K. Merton [17–18]. The process of obliteration or palimpsest (the latter expression referring to a piece of parchment used more than once, that is, being erased to make room for newer work) affects seminal works (i.e., truly ground-breaking research) offering novel ideas that are rapidly absorbed into the body of scientific knowledge. Such work is soon integrated into textbooks and becomes increasingly familiar within the scientific community. As a result of this absorption and canonization, the original sources (mainly articles or books) fail to be cited, either as full references (formal citations) or even as names or subject-specific terms (informal citations).

The ideas survive, sometimes becoming substantial elements of the basis and groundwork of modern science. But building over the groundwork implies obliteration of the sources. For example, the articles by Albert Einstein on the theory of relativity (published 1905 and 1916, respectively) are rarely cited in current research articles (as compared to less fundamental work) although they are the basis of modern cosmology and mainly caused Einstein's popularity. It may even happen that a transmitter, being familiar with the origin of a concept — and assuming the same of his readers — brings the idea back to life without citing the source, with the eventual result of becoming identified with its originator.

Eugene Garfield, the inventor of the citation indexes and the founder of ISI (Institute for Scientific Information, Philadelphia), concisely stated in one of his essays [19]: "Obliteration — perhaps even more than an astronomical citation rate — is one of the highest compliments the community of scientists can pay to the author.... It would mean that his contribution was so basic, so vital, and so well-known that scientists everywhere simply take it for granted. He would have been obliterated into immortality." Bearing this in mind, we should not expect that formal or even informal citations of the works of Kirchhoff can be taken as a real measure of the influence of his ideas in modern science. There are no metrics for quantifying fundamentality, significance or even elegance, which are terms falling under a completely different category.

References

[1] Thomson Reuters: http://scientific.thomsonreuters.com/products/wos/

[2] STN International Karlsruhe: http://www.stn-international.de/

[3] Marx, Werner & Manuel Cardona: The citation impact outside references — Formal versus informal citations, *Scientometrics* **80**, 1 (2009): 1–21. arXiv: physics/0701135v1

[4] Wu, F.Y.: The Potts model, *Reviews of Modern Physics* 54 (1982): 235–268.

[5] Kirchhoff, Gustav Robert: Ueber die Auflösung der Gleichungen, auf welche man bei der Untersuchung der linearen Vertheilung galvanischer Ströme geführt wird, *Annalen der Physik* 148 (2nd ser.) 72 (1847): 497–508.

[6] Kirchhoff, G.R: Zur Theorie der Lichtstrahlen, *Sitzungsberichte der Königlich Preußischen Akademie der Wissenschaften zu Berlin* 1882, part II: 641–669.

[7] Kirchhoff, G.R: Zur Theorie der Lichtstrahlen, *Annalen der Physik* 254 (3rd ser.) 18 (1883): 663–695.

[8] Kirchhoff, G.: Sur la théorie des rayons lumineux, *Annales Scientifiques de l'École Normale Supérieure* (3rd ser.) 3 (1886): 303–342.

[9] Graetz, Leo: Ueber die Wärmeleitfähigkeit von Flüssigkeiten, *Annalen der Physik* 254 (3rd ser.) 18 (1883): 79–94.

[10] Darrigol, Olivier: From organ pipes to atmospheric motions: Helmholtz on fluid mechanics, *Historical Studies in the Physical and Biological Sciences* **29**, 1 (1998): 1–51.

[11] Cruse, T.A. & F.J. Rizzo: A direct formulation and numerical solution of general transient elastodynamic problems (1), *Journal of Mathematical Analysis and Applications* 22 (1968): 244–259.

[12] SAO/NASA Astrophysics Data System: http://www.adsabs.harvard.edu/

[13] Physical Review Online Archive (PROLA): http://prola.aps.org/

[14] van Raan, A.F.J.: Sleeping beauties in science, *Scientometrics* 59 (2004): 461–466.

[15] Das, A. & G. Wichmann: Some extended analyses concerning the physics and kinematics of wave propagation in moving systems, *Zeitschrift für Angewandte Mathematik und Physik* 59 (2008): 156–180.

[16] Behrens, H.: Wissenschaftswachstum in wichtigen naturwissenschaftlichen Disziplinen vom 17. bis zum 21. Jahrhundert, *Berichte zur Wissenschaftsgeschichte* 29 (2006): 89–108.

[17] Merton, Robert K.: *Social Theory and Social Structure*, New York: The Free Press, 1968 (1st ed. 1949).

[18] Merton, R.K.: *On the Shoulders of Giants: A Shandean Postscript*, New York: The Free Press, 1965.

[19] Garfield, Eugene: The obliteration phenomenon in science — and the advantage of being obliterated, *Essays of an Information Scientist* 2 (1975): 396–398. http://www.thomsonreuters.com/business_units/scientific/free/essays/

[20] Maggi, Gian Antonio: Sulla propagazione libera e perturbata delle onde luminose in un mezzo isotropo, *Annali di Mathematica* (2nd ser) 16 (1888): 21–48.

[21] M.A.K. Hamid: Diffraction by a conical horn. *IEEE Transactions on Antennas and Propagation* AP16 (1968): 520–528.

Dr. Werner Marx, Scientist
Max Planck Institute for Solid State Research
Heisenbergstrasse 1
D-70569 Stuttgart (Germany)
Email: w.marx@fkf.mpg.de

Name Index

This index lists the names of all persons mentioned or quoted in the narrative aside from Gustav Robert Kirchhoff who is omitted. Composite terms incorporating proper names (e.g., Max Planck Society) are excluded. Italicized emphasis indicates that biographical information (living dates) is provided on the relevant page. An f. following a page number signifies that and the following page; ff. includes the subsequent page. For ranges exceeding three successive pages, the first and last pages are indicated. Initials occur only where a duplicate surname is separately cited.

d'Alembert, J.B., 65
Ångström, J.A., 11

Baker, B., 73, 119
Bessel, F.W., 3
Bessel-Hagen, E., 1
Bodenstein, M., 6
Bohr, N., 141
Boltzmann, L., 1–2, *7*, 13
Borchardt, K.W., 32, 34
Borchardt, W., 7
Born, M., 97, *112*, 128, 136–137
Borscheid, P., 6
Brömmel, L., 7
Brooker, G., 125, 132–136
Buchwald, J.Z., 63–124
Büttner, J., 18
Bunsen, R.W., *6*, 8–10, 15
Butrica, A.J., 15

Cauchy, A.L., 5, *71*
Chakrawartty, A., 126
Charpentier, E., 90, 120

Cheng, A.H.D. & D.T., 77, 120
Chladni, E., *5*
Clebsch, A., 32
Copson, E.T., 73, 119
Crookes, W., *15*

Darrigol, O., 90, 100, 120
Davy, H., 8
Debye, P., 27
Dieudonné, J., 120
Dirac, P.A.M., 80
Dirichlet, P.G.L., 40
Doppler, C., 16
Dove, H.W., 6

Einstein, A., 12, 150
Ettinghausen, A.v., 6
Euler, L., 5

Feyerabend, P., 129
Fitzpatrick, S., 126
Fourier, J., 73
Fraunhofer, J., *10*

Fresnel, A.J., 22–23, 31, 63–*64*, 72, 75, 116
Fröhlich, J., 31, 56

Galerkin, B., 5
Gander, M., 5
Garfield, 150, 151
Gauß, C.F., 6
Gödel, K., 128
Goldbach, Chr., 128
Goodman, J.W., 84, 120
Gouy, L.G., 100, 120
Gray, 90, 120
Green, G., 21, *75*, 79, 82, 89, 94–96, 101–104, 107, 118

Helmholtz, H.v., 22, 34, 79, 83, 97
Hensel, K., 1, 20, *80*
Hentschel, A.M., 31–62, 84
Hentschel, K., 1–18, 66
Herschel, J., *69*
Hertz, H., 1, 4, 14, 64, *105*
Hesse, L.O., 3
Heurtley, J.C., 127, 138
Hoffmann, D., 18
Hoffmann, K., 16
Hübner, K., 1
Huggins, W., *16*
Huygens, C., 22–24, 31, 34, 63, *66*, 75, 77, 89, 112, 116, 118

Jacobi, C.G.J., 3
James, F., 6
Jungnickel, C., 5–7, 13, 77, 121

Kamerlingh-Onnes, H., 1
Karpenko, V., 10
Kelvin, Lord → Thomson, William
Kipnis, N., 67, 121
Kirsten, C., 7
Klein, F., *14*–15, *100*
Kline, M., 77, 121
Königsberger, Leo, *7*
Körber, H.G., 7
Kong, J.A., 69, 121
Kottler, F., 107, *112*–113

Kragh, H., 127
Kuhn, T.S., 18, 64, 121
Kurlbaum, F., *17*–18

Landé, A., 100
Landstorfer, F.M., 19
Lang, V.v., 1
Laue, M.v., 100
Lecoq de Boisbaudran, P.E.F., *15*
Lippmann, G.J., 1
Lockemann, G., 6
Lockyer, J.N., *15*
Lorentz, H.A., 76–77, 121
Lucke, R.L., 113, 121
Lummer, O., *17*

MacDonald, H.M., 19–20, 29
Mach, E., 14
Maggi, G.A., 25, 29, *105*–8, 110–2, 121, 147, 151
Magnus, G., 6
Marchand, E.W., 63, *97*, 112–113, 117, 121
Marx, W., 66, 143–151
Maxwell, J.C., 4, 13, 21, 28, *72*
McCormmach, R., 5–7, 13, 77, 121
Merton, R., 149, 151
Meyenn, K.v., 1
Michelson, A.A., 117

Neumann, F.E., *3*, 6–7, *77*, 97

Ohm, G.S., *4*
Olesko, K.M., 3
Ostwald, 1, 14

Pais, A., 141
Pauling, L., 100
Pease, F.G., 117
Peierls, R., 100
Planck, M., 1, 12, *16*, 18
Poincaré, H., 63, *65*, 90–96, 100, 103–105, 111, 122
Poisson, S.D., 73
Popper, K.R., 128
Pringsheim, E., *16*

Rabi, I., 100
Ramsey, W., *15*
Rayleigh, Lord → Strutt, W.
Raynaud, J., *15*
Richelot, C., 5
Richelot, F.J., 3
Riemann, B., *75*
Ritz, W., *5*
Roscoe, H.E., *8*–9
Rubens, H., *16*
Rubinowicz, A., 25, 29–30, 63,
    107–112
Runge, I., 27, 30

Saatsi, J., 66, 126
Schellen, H., 15
Schirrmacher, A., 12
Schöpf, H.G., 18
Schuster, A., 1, 7
Seth, S., 100, 121
Shapiro, A., 67
Siegel, D.M., 12
Smith, P., 128
Sommerfeld, A., 22, 25, 27, 30, *65*,
    97, *100*–104, 107, 116–117, 127
Stewart, B., 12

Stokes, G.G., *11*, 70–*71*, 73–77,
    122–123
Strutt, W. (Lord Rayleigh), *65*,
    96–99, 103, 116–117, 127

Thomson, W. (Lord Kelvin), 5, *11*, *75*
Todhunter, I., 5

Ufimtsev, P.Y., 20, 30, 137

Vickers, P., 66, 125–142
Vogel, H.C., *16*
Voigt, W., 31

Weber, W., 6
Weger, A., 17
Weingartner, P., 129
Wiedemann, Eilhart, 1, 31, 56
Wolf, E., 63, *97*, 112–113, 117, 121,
    128, 136–137
Wollaston, W.H., *10*

Yeang, C.-P., 63–124
Young, T., *67*, 100, 111

Zhu, N.Y., 19–30, 66, 80, 137